Interleukins - The Immune and Non-Immune Systems' Related Cytokines

Edited by Payam Behzadi

Published in London, United Kingdom

IntechOpen

Supporting open minds since 2005

Interleukins - The Immune and Non-Immune Systems' Related Cytokines
http://dx.doi.org/10.5772/intechopen.92961
Edited by Payam Behzadi

Contributors
Fauziyya Muhammad, Ummey Khalecha Bintha Ahmed, Souwelimatou Amani, Md Huzzatul Mursalin, Fortunato Vesce, Samuel Evans Adunyah, Richard Akomeah, Fareed K.N. Arthur, Roland S. Cooper, Joshua C. M. Williams, Gregory D. Maniero, Yogita Sharma, Neeraj Kumar, Devyani Thakur, Daniel J. Hellenbrand, Carolyn N. Morehouse, Rylie M. Roddik, Sophia M. Mauney, Ryan T. Elder, Amgad S Hanna, Payam Behzadi, Márió Gajdács, Herney Andrés Andrés García-Perdomo, Meysam Sarshar, Daniela Scribano, Cecilia Ambrosi

First published in London, United Kingdom, 2021 by IntechOpen
IntechOpen is the global imprint of INTECHOPEN LIMITED, registered in England and Wales, registration number: 11086078, 5 Princes Gate Court, London, SW7 2QJ, United Kingdom
Printed in Croatia

British Library Cataloguing-in-Publication Data
A catalogue record for this book is available from the British Library

Additional hard and PDF copies can be obtained from orders@intechopen.com

Interleukins - The Immune and Non-Immune Systems' Related Cytokines
Edited by Payam Behzadi
p. cm.
Print ISBN 978-1-83969-098-3
Online ISBN 978-1-83969-099-0
eBook (PDF) ISBN 978-1-83969-100-3

For EU product safety concerns:
IN TECH d.o.o., Prolaz Marije Krucifikse Kozulić 3, 51000 Rijeka, Croatia, info@intechopen.com or visit our website at intechopen.com.

Meet the editor

Dr. Payam Behzadi was born in Tehran, Iran, in 1973. He began his collaboration with the Department of Microbiology, College of Basic Sciences, Shahr-e-Qods Branch, Islamic Azad University as a faculty member in 2004. He has a BSc and MSc in Microbiology and a Ph.D. in Molecular Biology and now continues his scientific activities in the position of assistant professor at Islamic Azad University. He has authored and edited more than twenty chapters and academic books and more than seventy original and review articles. His scientific research interests include urinary tract infections, antibiotics, bioinformatics, genetics, gene profiling, molecular biology, and cellular and molecular immunology. Dr. Behzadi trains as an ice skater in his free time.

Contents

Preface

Interleukins, as a collection of cytokines with a pivotal role in immune and non-immune systems, are spread across the human body. These multi-functional glycoproteins support the human body in different conditions and situations, for example, in health and in diseases.

Indeed, interleukins have different functions, characteristics, and structures. Hence, these diverse biomolecules are classified into different families and subfamilies.

Interleukins have different patterns of behavior in different conditions. These immune glycoproteins act as variable molecules in health and diseases. Their activities depend on the related conditions and act in a cascade pattern. This characteristic of interleukins makes them invaluable biomarkers in diagnostics. Therefore, in recent years, interleukins are recognized as effective options for prognosis, diagnosis, and control of related conditions in the human body.

This book is the outcome of international collaboration, cooperation, and teamwork between a collection of international scientists around the world.

It is divided into three main sections: "Interleukins' Classification and Evolutionary Features", "Autoimmune Diseases and Low Immune System," and "Cancer and Injuries".

The first section includes two chapters: "A World of Wonders: Interleukin-1 (IL-1) and IL-2 Families" and "Evolutionary Conservation of the Role of CD4 as a Receptor for Interleukin-16". These chapters reveal the classification and evolutionary features regarding their topics.

Section two includes three chapters: "Interleukin 6 in Patients with Rheumatoid Arthritis", "Therapeutic Potential of IL-9 in Allergic and Autoimmune Diseases", and "From Pregnancy Loss to COVID 19 Cytokine Storm: A Matter of Inflammation and Coagulation". These chapters provide invaluable scientific information regarding autoimmune diseases and the role of interleukins, weakened immune systems in pregnant women, and COVID-19 and related interleukins.

Section three includes two chapters: "IL-17 Biological Effects and Signaling Mechanisms in Human Leukemia U937 Cells" and "The Role of Interleukins after Spinal Cord Injury."

Interleukins - The Immune and Non-Immune Systems' Related Cytokines is the outcome of sincere and effective collaboration between the editor, chapter authors, and publisher, all of whom actively and effectively contributed to the production of this scientific reference.

I wish to thank my great colleagues including Dr. Márió Gajdács (from Hungary), Prof. Andrés García-Perdomo (from Colombia), Dr. Meysam Sarshar (from Italy),

Dr. Daniela Scribano (from Italy) and Prof. Cecilia Ambrosi (from Italy) who supported me to prepare the first chapter of this book by their brilliant collaboration and cooperation.

And finally, I appreciate Author Service Manager Ms. Sara Debeuc at IntechOpen for her great support and cooperation. In addition, I am grateful to Lucija Tomicic-Dromgool, Martina Usljebrka Kauric, and Anja Filipovic, commissioning editors at IntechOpen for their excellent collaboration, management, and arrangement for preparing this valuable book.

Payam Behzadi
Assistant Professor,
Department of Microbiology,
College of Basic Sciences,
Shahr-e-Qods Branch,
Islamic Azad University,
Tehran, Iran

Section 1

Interleukins' Classification and Evolutionary Features

Chapter 1

A World of Wonders: Interleukin-1 (IL-1) and IL-2 Families

Márió Gajdács, Herney Andrés García-Perdomo, Meysam Sarshar, Daniela Scribano, Cecilia Ambrosi and Payam Behzadi

Abstract

Human interleukins (ILs) are a collection of different biological molecules belonging to the group of cytokines, associated with various immune and non-immune systems and different signaling pathways. ILs contribute to the function of different tissues, organs and systems in the human body. They are involved in homeostasis, infectious diseases, autoimmune diseases, cancers and even therapeutics. Due to this knowledge, this chapter aims to summarize the importance of the IL-1 and IL-2 superfamilies.

Keywords: immune system, cytokines, interleukins, interleukin-1, interleukin-2

1. Introduction

Human cytokines are consisted of a wide range of proteins (and/or glycoproteins) known as immune molecules with different properties that affect both immune and non-immune cells. Based on different characteristics, cytokines are classified into several groups including interleukins (ILs), interferons (IFNs), chemokines and lymphokines [1–4].

In 1979 at the cytokinologists' meeting of "the Second International Lymphokine Workshop" in Switzerland, the term of *Interleukin* was officially proposed for the first time. The proposed term of "*Interleukin*" was published in a letter to the editor by the *Journal of Immunology* [2, 5–7].

However, this term is not entirely correct, because the ILs are not only limited to leukocytes but they also involve other cells other than leukocytes [4]. The majority of IL glycoproteins are produced by endothelial cells, monocytes, macrophages (MΦs) and T helper (CD4+) lymphocytes [3].

There are several nomenclature systems, which may be applied for the categorization of ILs. However, interleukins are recognized through the capital letters IL followed immediately by a dash and a number e.g. IL-17 [3, 5, 8]. One of these nomenclature systems was approved by the subcommittee of the nomenclature committee of the International Union of Immunological Societies (IUIS) and the World Health Organization (WHO). Functional characteristics, structural properties, amino acid sequences and related homology, types of receptors (among other things) may be recognized as important criteria for ILs' classification [2, 9].

The authors try to discuss general characteristics, structures, classifications, and genomic maps of ILs throughout this chapter.

2. General characteristics and structure of interleukins

ILs as a group of cytokines, which are involved in immune and non-immune cell activation, cell adhesion, cell differentiation, cell maturation and cell migration; in other words, these proteins act like signaling molecules that induce different pathways in the human body. Although ILs encompass a wide range of functions and structures, they participate in immunomodulatory activities and also contribute to pro-inflammatory and anti-inflammatory responses. These processes initiate through the attachment of the IL biomolecules to their specific receptors onto the cells, which may lead to induction of immune responses. However, the efficacy and specificity of these responses is associated with the related receptors, ligands and signaling pathways [2, 3, 10, 11].

In addition, IL proteins have pivotal role in cancers. Indeed, these biomolecules are produced and secreted by tumor- and immune cells within the tumors. Due to these facts, ILs affect the processes of angiogenesis, invasion, growth and immune responses related to tumors. Because of the presence of ILs in different cells, tissues and organs, they have recognized as invaluable biomarkers in diagnostics and therapeutic planning [11, 12].

Up to date, more than 40 ILs have been identified with a wide range of subtypes. The structural characteristics of cytokines, including IL proteins, are effective criteria for their categorization. In this regard, some IL glycoproteins are divided into two groups: type I (class I) and type II (class II) cytokines. Type I cytokines bear a general structure comprising four compact α-helices within tensed packages. The arrangement of the related α-helices involves a four-helix bundle with an antiparallel (up-up-down-down) configuration. Type II cytokines obey the same structure as described for type I cytokines. However, type II cytokines bear the compact packages of six to seven-helix bundles with the configuration of antiparallel arrangement [2, 3, 11, 13–15].

Furthermore, in accordance with the length of bundles made of α-helices, type I cytokines are classified into short- and long-chains subclasses. The members of long-chain subclass pertaining to type I cytokines encompass bundle peptides with more than 165 amino acids, while the members of short-chain subclass belonging to type I cytokines possess bundle peptides with less than 165 amino acids [3, 16]. Interleukins are classified into the families of IL-1 family, IL-2 family, IL-6 family (including ILs 6, 11, 31, cardiotrophin-like cytokine (CLC), cardiotrophin-1 (CT-1), ciliary neurotrophic factor (CTNF), leukemia inhibitory factor (LIF), and oncostatin M (OSM)) [17]; IL-10 family (composed of ILs 10,19, 20, 22, 24, 26) [18]; IL-12 superfamily (comprising ILs 12, 23, 27, 35) [19]; and IL-17 family (containing ILs 17 A-D and IL-25 (IL-17E)) [20]. IL-8 belongs to the CXC-chemokines and is classified along with them [21].

3. The IL-1 superfamily and the related members

IL-1 superfamily members (IL1-like cytokines) involve functionally quite distinct molecules composed of IL-37, as the single member anti-inflammatory cytokine, IL-1Ra, IL-36Ra and IL-38 as three well-known receptor blockers or antagonists, and IL-1α, IL-1β, IL-18, IL-33, IL-36α, IL-36β, IL-36γ as seven ligands with agonistic functions [2, 3, 22, 23]. The IL-1 is known as the pioneer member of the IL-1 superfamily and its receptor was recognized as IL-1R. Interestingly, the IL-1R encompasses a molecule of Toll-IL-1 receptor (TIR) domain in its structure. The TIR domain – which is identified in both structures of IL-1R and toll-like receptor (TLR) glycoproteins – has pivotal roles in transduction of internal signals by different stimulators e.g. recognition of pathogen/microbe-associated molecular patterns (PAMPs/MAMPs) and danger/damage-associated molecular patterns (DAMPs) by ectodomain structures of TLRs to strengthen the immune responses and inflammation [22, 24–27]. The gene clusters of IL-1 superfamily members, excluding ILs 18 and – 33, map within chromosome 2 in humans (**Table 1**) [2, 3, 22].

IL was identified for the first time in 1979. IL-1α and IL-1β proteins (**Figure 1**) have the same biological characteristics but the lowest homology in their sequences. Moreover, the nature of IL-1α subunit makes it active with effective biological functions, while the IL-1β is produced as a pro-subunit which the enzyme of caspase-1 should be activated [3, 12, 22, 29, 34].

The IL-1 superfamily members are produced by different of immune and non-immune cells such as chondrocytes, dendritic cells (DCs), epithelial- and endothelial cells, keratinocytes, lymphocytes, fibroblasts, MΦs, monocytes, neutrophils and smooth muscle cells [2, 35–37].

As presented on **Figures 1** and **2** each subunit whether IL-1α or IL-1β, encompasses 11 loops and 12 β-strands which appear as a typical configuration of β-trefoil. The two ILs 1α and 1β have no specific binding to the IL-1 receptor type I and II, respectively; they both can bind both types, but, in case of type I receptor (CD 121a), with stimulatory action and pathway signaling, and, in case of the type II receptor, (decoy receptor that lacks the TIR domain) with inhibitory effect similar to the IL1Ra binding. The connections between the ligands and receptors are supported by the presence of IL-1R accessory co-receptor proteins (IL-1RAcP) to prepare proper conformational changes [3, 22].

IL-1 superfamily member		Gene		Active form	Molecular weight (active structure)	Chromosome	References
IL-1α/IL-1F1		*IL-1α/IL1F1*		Heterodimer; β-trefoil fold	17 kDa	2q14	[2, 22, 28, 29]
IL-1β/IL-1F2		*IL-1β/IL-1F2*			17 kDa	2q14	[2, 22, 28, 29]
IL-1Ra/IL-1F3		*IL-1Ra/IL-1F3*			22 kDa	2q14.2	[2, 22, 28, 29]
IL-18/IL-1F4		*IL-18/IL-1F4*		Homodimer; β-trefoil fold	18 kDa	11q22.2-11q22.3	[2, 22, 28, 29]
IL-33/IL-1F11		*IL-33/IL-1F11*		β-trefoil fold	18 kDa	9p24.1	[2, 22, 28, 29]
IL-36/ IL-1F	α/6	*IL-36/ IL-1F*	*A/6*	β-trefoil fold	35 kDa	2q12-2q14.1	[2, 22, 23, 28]
	β/8		*B/8*			2q14	
	γ/9		*G/9*			2q12-2q21	
	Ra/5		*Ra/5*		17 kDa	2q14	[2, 22, 28, 30]
IL-37/IL-1F7		*IL-37/IL-1F7*		β-trefoil fold	17–24 kDa	2q12-2q14.1	[2, 22, 28, 31]
IL-38/IL-1F10		*IL-38/IL-1F10*		β-trefoil fold	17 kDa	2q14	[2, 22, 28, 32]

Table 1.
IL-1 superfamily members, related genes, molecular weight and loci.

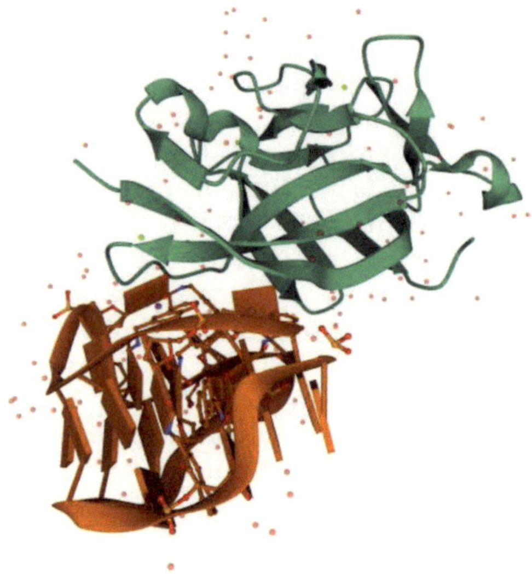

Figure 1.
The IL-1α structure including an α-helix, β-sheets and loops (shown in green color) (PDB ID 5UC6) [33].

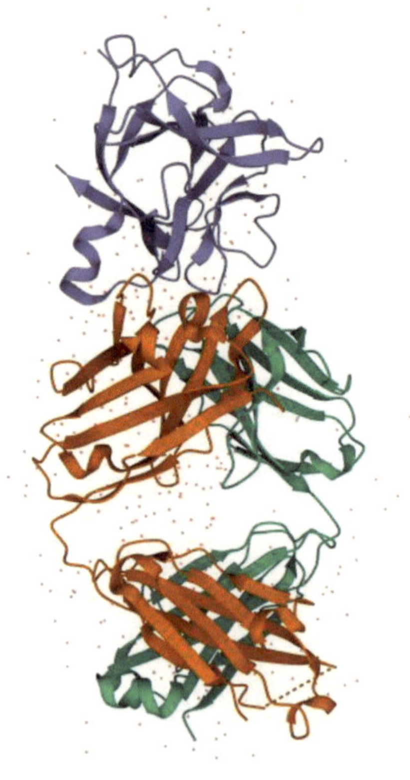

Figure 2.
The IL-1β structure including an α-helix, β-sheets and loops (shown in violet color) (PDB ID 7CHZ) [38].

The heterodimers of IL-1α and IL-1β activate the inflammatory pathway via the employment of MyD88, which may lead to expression of the nuclear factor κB (NF-κB) and consequently, to expression of inflammatory products. Thus, the

IL-1 is known as endogenous pyrogen. The IL-1 functional properties are modulated by IL-1Ra. However, the IL-1RAcP binding domain is absent in IL-1Ra. Thus, the IL-1 signaling pathway cannot be induced via the IL-1Ra molecule [3, 22, 29, 39].

IL-1 is also involved in tumor biology through the processes of angiogenesis, invasion and proliferation. The expression of IL-1 proteins increase within a wide range of cancers e.g., melanoma, lung, breast, neck, colon, head among others [11, 40].

IL-18 is another member of IL-1 superfamily, which was identified in 1989 and introduced as the triggering factor of interferon-γ (IFN-γ). IL-18, resembling IL-1β is produced as an inactive protein and should be activated by the cleavage done via caspase-1. IL-18 is the ligand of the double stranded and heterodimeric receptor of IL-18R complex. The combined complex of IL-12 and IL-18 may lead to significant expression of IFN-γ. The expression of IFN-γ resulted from IL-18 and IL-12 combination has anti-tumoral effects. The expression of IL-18 increases in breast-, hepatocellular-, lung-, esophageal-, ovarian- and renal carcinomas [2, 11, 22, 29, 34, 41–43]. The immune and non-immune cells (e.g. astrocytes, DCs, keratinocytes, Kupffer-cells, MΦs and osteoblasts) are important resources for IL-18 expression and secretion [29, 44].

IL-33 as another IL-1 superfamily member contributes to type 2 immunity and inflammation. Indeed, the IL-33/ST2 pathway has pivotal roles in activation of type 2 immunity through triggering of those T helper 2 (Th2) cells which produce suppression of tumorigenicity 2 (ST2) molecules [3, 45–47]. The main resource cells that produce IL-33 include keratinocytes, mucosal tissues, fibroblasts, endothelial- and epithelial cells [3, 22, 45, 48]. IL-33 binds to the related receptor of IL-1RL1 (or ST2). Furthermore, the related co-receptor is IL-1RAcP [45]. The precursor of the alarmin cytokine IL-33 is inactivated by caspase-1, while the enzymes of neutrophil elastase and serine proteases cathepsin G cleave the IL-33 precursor into active mature structures. However, the 30 kDa precursor of IL-33 is functional upon necrotic cell death and cell damage [3, 22, 34, 45, 49]. The IL-33 receptor, ST2 mediates activation of the MyD88-dependent signaling pathway. The ST2 molecules are produced by different innate and adaptive immune cells. The ST2 molecule participates in type 2 cytokines expressions which are involved in inflammatory and allergic responses [3, 22, 29]. IL-33 supports the homeostasis of intestinal microbiota through the induction of Immunoglobulin A (IgA) in adaptive immune B cells. In addition, the IL-33 receptor of ST2 is capable to inhibit the colorectal cancer via reduction of regulatory T cells (T_{reg}) infiltration and enhancement of CD8+ T cells [12, 45, 50].

IL-36 cytokines belong to the subfamily members of IL-1 superfamily. The subfamily of IL-36 is comprised of three IL-36 receptor (IL-36R) agonists of IL-36α, IL-36β, and IL-36γ and an IL-36R antagonist of IL-36Ra. The IL-36α, IL-36β, and IL-36γ have 24%, 27% and 20% similar homology with IL-1Ra, respectively and 30%, 31% and 31% similarity with IL-1β, respectively [23, 51, 52]. The IL-36α, IL-36β, and IL-36γ bind to IL-36R through the co-receptor IL-1RAcP. The IL-36α, IL-36β, and IL-36γ are expressed as biologically inactive precursors which should be activated by proteases of elastase/Cat G, Cat G, proteinase3/elastase/CatG, respectively [23]. The IL-36α, IL-36β, and IL-36γ are recognized as inflammatory cytokines which activate the related signaling pathway via activation of NF-κB, while the antagonist of IL-36Ra as the anti-inflammatory cytokine inhibits the inflammatory signaling pathway through inactivation of L-36R [23, 53]. The expression of IL-36γ cytokine is observed in cancers of colorectal, esophageal, melanoma, lung and neck and squamous cell carcinoma [11, 54].

IL-37, which is the anti-inflammatory member of the IL-1 superfamily was identified in 2000. IL-37 cytokines consist of five isoforms including IL-37a (isoform 5), IL-37b (isoform 1), IL-37c (isoform 4), IL-37d (isoform 2) (expressed only in the testes and the bone marrow) and IL-37e (isoform 3) (expressed only in the testes and the bone marrow). The isoform of IL-37a is the only that is expressed in the brain (along with the heart and kidney, like the IL-37b and IL-37c, –and also in lymph nodes, bone marrow, placenta, lung-; IL-37a is brain-specific between the three), IL-37b is kidney-specific between the three, and, finally, IL-37c is heart-specific (the only of them expressed in the heart). Moreover, the IL-37b or isoform 1 is the largest member of the IL-37 isoforms with a length of 218 amino acids. The IL-37 isoforms are produced as precursors which should be biologically activated by protease enzymes e.g. caspase-1 [55, 56]. The IL-37b has considerable sequence similarity with IL-18 and therefore the isoform 2 of IL-37 is capable to bind to the α-chain of IL-18 receptor (IL-18Rα) [29, 55]. The IL-37 prevents the progression of colon cancer through inhibition of β-catenin [12].

IL-38 is known as bioinformatic cytokine, because it was detected by *in silico* studies. This cytokine was discovered in 2001, and has 41% sequence similarity with IL-1Ra, 43% with IL-36Ra and 29% with IL-1β. Interestingly similar to IL-36Ra, the IL-38 behaves as an antagonist and binds to IL-36R [3, 22, 57]. The *IL-38* gene maps to chromosome 2, situated within a gene cluster containing *IL-1α-IL-1β-IL-37-IL-36γ-IL-36α-IL-36β-IL-36Ra-IL-38-IL-1Ra* that spans from the centromere to the telomere [57]. In accordance with previous studies, IL-38 enhances the pro-inflammatory immune responses against LPS and inhibits the expression of IL-17 triggered by *Candida albicans* and IL-22 [11, 22, 57–59].

4. The IL-2 superfamily and related members

The IL-2 superfamily is known as the γ_c family of cytokines [2, 3, 29, 60]. The IL-2 superfamily or γ_c family of cytokines comprise six members including IL-2, IL-4, IL-7, IL-9, IL-15 and IL-21 (**Table 2**); the members of this superfamily are classified as type I cytokines bearing four α-helical bundle. γ_c (CD132) is the pivotal protein portion of IL-2, IL-4, IL-7, IL-9, IL-15 and IL-21 receptors. The first recognized member of this family, IL-2 was identified in 1976 as a T-cell growth factor [3, 60–62]. The IL-2 superfamily members contribute to a wide range of immune cells' activities such as regulation of B cell development, innate lymphoid cells, natural killer (NK) and T cells, differentiation, proliferation and survival of immune cells [60].

The IL-2 cytokine is expressed by a wide range of innate and adaptive immune cells, including DCs, NK cells, NK T (NKT) cells, active $CD4^+$ and $CD8^+$ T cells. Interestingly, the IL-2 cytokine targets the B immune cells, NK and $CD4^+$ and $CD8^+$ T. IL-2Rα (known as T_{ac} antigen or CD25), IL-2Rβ (known as CD122) and IL-2Rγ (known as CD132 or γ_c) together form a three-segmented receptor for IL-2

IL-2 superfamily member	Gene	Active form	Molecular weight (active structure)	Chromosome	References
IL-2	*IL-2*	monomer	15.5 kDa	4q26-4q27	[2, 29]
IL-4	*IL-4*	monomer	15 kDa	5q23-5q31	[2, 29]
IL-7	*IL-7*	monomer	25 kDa	8q12-8q13	[2, 29]
IL-9	*IL-9*	monomer	14 kDa	5q31-5q35	[2, 29]
IL-15	*IL-15*	monomer	14–15 kDa	4q31	[2, 29]
IL-21	*IL-21*	monomer	15 kDa	4q26-4q27	[2, 29]

Table 2.
IL-2 superfamily members, related genes, molecular weight and loci.

[3, 29, 60, 61]. The IL-2 cytokine has therapeutic application for some cancers, e.g., melanoma and renal carcinoma. IL-2 alone or together with related vaccines and/or cytokines can be considered as therapeutic options [11, 63].

IL-4, another member of the IL-2 superfamily, is secreted by activated basophils, eosinophils, mast cells, NKT cells, T helper 2 (T_{h2}) cells and γδ T cells [29, 64]. It participates in the differentiation of T_{h2} and T_{h9}, mediates allergic conditions, triggers expression of IgE in B cells, and protects the human body from infectious diseases caused by extracellular parasites and helminths. However, it was shown that rather than IL-4, the ILs of 1, 2, 25, and 33 enhance the process of T_{h9} differentiation [60, 65]. IL-4 and IL-13 – with 25% sequence similarity – cooperate with each other in some functional activities [3, 29, 60, 64].

IL-4 has 2 groups of receptors: type I (composed of IL-4Rα or CD124 and the γ_c or CD132) and type II (involves IL-4Rα and IL-13Rα1). The type I IL-4R is able to bind only to IL-4 while the type II IL-4R is capable to bind to the both of IL-4 and IL-13 [29, 66]. High levels of IL-4Rs are detected in tumors. The IL-4 participates in tumorigenesis through prevention of T_{h1} activation and activation of T_{h2}. The expression of IL-4 increases in different types of cancers including urinary bladder, breast, colon, lung ovary, pancreas and the prostate [11, 67].

IL-7 is a critical cytokine in B- and T-lymphocyte development and maturation. The main expression resources of IL-7 are immune and non-immune cells, such as B cells, DCs, epithelial cells, keratinocytes, MΦs and monocytes. IL-7 is the ligand of the complex structured IL-7R which is composed of IL-7Rα or CD127 and the γ_c or CD132 [29, 66]. IL-7 is significantly involved in homeostatic regulation of the both groups of immature and mature T cell types [2, 60]. In addition, IL-7 triggers the anti-tumoral immune responses and prevents the growth of tumors [11].

IL-9 is expressed by the immune cells of eosinophils, mast cells and T cells of T_{h2} and T_{h9}. IL-9 triggers the expression of IgE by the adaptive B immune cells and the secretion of mucosal production and chemokines in the bronchi. IL-9 is involved in helminth infectious diseases, allergies and asthma. The IL-9R is composed of two subunits of IL-9Rα and the γ_c or CD132 [3, 29]. Moreover, the IL-9 acts as a double-edged sword. In melanoma, IL-9 acts as tumor inhibitor while in acute leukemia, it acts as a tumorigenetic cytokine [11].

IL-15 is expressed by a wide range of immune and non-immune cells consisting CD4+ T cells (functional), monocytes, keratinocytes and skeletal muscle cells. IL-15 and IL-2 have significant similarities in their structures. The IL-15R involves three subunits of IL-15Rα, IL-2Rβ and the γ_c or CD132. IL-15 participates in proliferation of NK cells and the homeostasis of memory $CD8^+$ T cells [3, 29, 60]. Additionally, IL-15 has anti-tumoral effects [11]. IL-15, unlike IL-2, does not stimulate LT_{reg}, that can be useful in cancer therapy. Treg stimulation is one of the negative sides of IL-2 that IL-15 does not share [68–70].

IL-21 is expressed by immune cells such as NKT, T and mainly by T_{h17} and T follicular helper (T_{fh}) [29, 60]. Similarly to IL-4, IL-21 enhances the expression of IgG1; on the other hand, conversely from IL-4, IL-21 prevents the expression of IgE [60, 71]. The IL-21 receptor is composed of two subunits of IL-21R and the γ_c or CD132. IL-21 has anti-tumoral effects on some groups of cancers which may lead to recuperate melanoma and renal carcinoma; while the IL-21 is tumorigenic in colorectal cancer [3, 11].

5. Conclusion

Cytokines involve a wide range of molecules with different structural characteristics, functional properties and biological activities. Identification and recognition of these characteristics help us to understand their classification and categorization. Each group of these biomolecules has its massive importance and huge application prospects. Indeed, ILs have pivotal roles in different parts of the human body, particularly associated with the innate and adaptive immune system. A complete understanding of ILs characteristics and properties will improve the successful outcomes against infectious and autoimmune diseases, as well as cancers.

Acknowledgements

The authors have special thanks to Sara Debeuc the author service manager, Lucija Tomicic-Dromgool and Martina Usljebrka Kauric the Commissioning Editors of InTechOpen Company for giving us this opportunity to prepare this chapter.

Funding

M.G. was supported by the János Bolyai Research Scholarship (BO/00144/20/5) of the Hungarian Academy of Sciences. M.S. and D.S. salaries were supported by Italian Ministry of Health (Progetto SG-2018-12365432) and the Dani Di Giò Foundation-Onlus, Rome, Italy, respectively. The research was supported by the ÚNKP-20-5-SZTE-330 New National Excellence Program of the Ministry for Innovation and Technology from the source of the National Research, Development and Innovation Fund. Support from the Ministry of Human Capacities, Hungary, grant 20391–3/2018/FEKUSTRAT is acknowledged. M.G. would also like to acknowledge the support of ESCMID's "30 under 30" Award. The funders did not play a role in drafting and writing this manuscript.

Conflict of interest

The authors declare no conflicts of interest.

Author details

Márió Gajdács[1,2], Herney Andrés García-Perdomo[3], Meysam Sarshar[4], Daniela Scribano[5,6], Cecilia Ambrosi[7] and Payam Behzadi[8*]

1 Faculty of Medicine, Institute of Medical Microbiology, Semmelweis University, Budapest, Hungary

2 Department of Pharmacodynamics and Biopharmacy, Faculty of Pharmacy, University of Szeged, Hungary

3 Division of Urology, Department of Surgery, School of Medicine, UROGIV Research Group, Universidad del Valle, Cali, Colombia

4 Research Laboratories, Bambino Gesù Children's Hospital, IRCCS, Rome, Italy

5 Department of Public Health and Infectious Diseases, Sapienza University of Rome, Rome, Italy

6 Dani Di Giò Foundation-Onlus, Rome, Italy

7 Department of Human Sciences and Promotion of the Quality of Life, San Raffaele Open University, IRCCS, Rome, Italy

8 Department of Microbiology, College of Basic Sciences, Shahr-e-Qods Branch, Islamic Azad University, Tehran, Iran

*Address all correspondence to: behzadipayam@yahoo.com

References

[1] Gajdács M, Behzadi P. Introductory Chapter: Cytokines-The Diamonds and Pearls of Biological Systems. In: Behzadi P, editor. Cytokines. Croatia: IntechOpen; 2020. p. 1-7.

[2] Brocker C, Thompson D, Matsumoto A, Nebert DW, Vasiliou V. Evolutionary divergence and functions of the human interleukin (IL) gene family. Human genomics. 2010;5(1):1-26.

[3] Ferreira VL, Borba HH, Bonetti AdF, Leonart LP, Pontarolo R. Cytokines and interferons: types and functions. Autoantibodies and cytokines: IntechOpen; 2018.

[4] O'Shea JJ, Gadina M, Siegel RM. Cytokines and cytokine receptors. Clinical immunology: Elsevier; 2019. p. 127-55. e1.

[5] Mizel SB, Farrar JJ. Revised nomenclature for antigen-nonspecific T-cell proliferation and helper factors. Cellular Immunology. 1979;48(2):433-436.

[6] Oppenheim JJ. Cytokines, their receptors and signals. The autoimmune diseases: Elsevier; 2020. p. 275-289.

[7] Cameron MJ, Kelvin DJ. Cytokines, chemokines and their receptors. Madame Curie Bioscience Database [Internet]: Landes Bioscience; 2013.

[8] di Giovine FS, Duff GW. Interleukin 1: the first interleukin. Immunology today. 1990;11:13-20.

[9] Paul W, Kishimoto T, Melchers F, Metcalf D, Mosmann T, Oppenheim J, et al. Nomenclature for secreted regulatory proteins of the immune system (interleukins). WHO-IUIS Nomenclature Subcommittee on Interleukin Designation. Clinical and experimental immunology. 1992;88(2):367.

[10] Commins SP, Borish L, Steinke JW. Immunologic messenger molecules: cytokines, interferons, and chemokines. Journal of Allergy and Clinical immunology. 2010;125(2):S53-S72.

[11] Farc O, Cristea V. PRO-AND ANTITUMOR ROLE OF THE INTERLEUKINS 1 TO 41. Romanian Archives of Microbiology and Immunology. 2019;78(3):149-162.

[12] Li J, Huang L, Zhao H, Yan Y, Lu J. The Role of Interleukins in Colorectal Cancer. International Journal of Biological Sciences. 2020;16(13):2323.

[13] Ealick SE, Cook WJ, Vijay-Kumar S, Carson M, Nagabhushan TL, Trotta PP, et al. Three-dimensional structure of recombinant human interferon-gamma. Science. 1991;252(5006):698-702.

[14] Zdanov A, Schalk-Hihi C, Gustchina A, Tsang M, Weatherbee J, Wlodawer A. Crystal structure of interleukin-10 reveals the functional dimer with an unexpected topological similarity to interferon γ. Structure. 1995;3(6):591-601.

[15] Kotenko SV. The family of IL-10-related cytokines and their receptors: related, but to what extent? Cytokine & growth factor reviews. 2002;13(3):223-240.

[16] Boulay J-L, O'Shea JJ, Paul WE. Molecular phylogeny within type I cytokines and their cognate receptors. Immunity. 2003;19(2):159-163.

[17] Rose-John S. Interleukin-6 family cytokines. Cold Spring Harbor perspectives in biology. 2018;10(2):a028415.

[18] Wei H, Li B, Sun A, Guo F. Interleukin-10 family cytokines immunobiology and structure. Structural Immunology. 2019:79-96.

[19] Behzadi P, Behzadi E, Ranjbar R. IL-12 family cytokines: general characteristics, pathogenic microorganisms, receptors, and signalling pathways. Acta microbiologica et immunologica Hungarica. 2016;63(1):1-25.

[20] Zhang X, Angkasekwinai P, Dong C, Tang H. Structure and function of interleukin-17 family cytokines. Protein & cell. 2011;2(1):26-40.

[21] Baggiolini M, Dewald B, Moser B. lnterleukin-8 and related chemotactic cytokines—CXC and CC chemokines. Advances in immunology. 1993;55:97-179.

[22] Garlanda C, Dinarello CA, Mantovani A. The interleukin-1 family: back to the future. Immunity. 2013;39(6):1003-1018.

[23] Queen D, Ediriweera C, Liu L. Function and regulation of IL-36 signaling in inflammatory diseases and cancer development. Frontiers in cell and developmental biology. 2019;7:317.

[24] Dinarello CA. Immunological and inflammatory functions of the interleukin-1 family. Annual review of immunology. 2009;27:519-550.

[25] Behzadi E, Behzadi P. The role of toll-like receptors (TLRs) in urinary tract infections (UTIs). Central European journal of urology. 2016;69(4):404.

[26] Behzadi P. The Role of Toll-Like Receptor (TLR) Polymorphisms in Urinary Bladder Cancer In: Sameer AS, Banday, Mujeeb Zafar, Nissar, Saniya, editor. Genetic Polymorphism and cancer susceptibility. I. Singapore: Springer; 2021. p. 279-315.

[27] KARAKU FN. CHAPTER SEVEN GLYCOLIPIDS AND IMMUNITY FATMA NUR KARAKUş. Studies in Glycolipids. 2021:93.

[28] IB M. Cytokines. In: Firestein GS BR, Gabriel SE, McInnes IB, O'Dell JR, editor. Kelley and Firestein's Textbook of Rheumatology. 10th ed: Elsevier; 2017. p. 396-407.

[29] Akdis M, Burgler S, Crameri R, Eiwegger T, Fujita H, Gomez E, et al. Interleukins, from 1 to 37, and interferon-γ: receptors, functions, and roles in diseases. Journal of allergy and clinical immunology. 2011;127(3):701-21. e70.

[30] Zhou L, Todorovic V, Kakavas S, Sielaff B, Medina L, Wang L, et al. Quantitative ligand and receptor binding studies reveal the mechanism of interleukin-36 (IL-36) pathway activation. Journal of Biological Chemistry. 2018;293(2):403-411.

[31] Khosh E, Bahmaie N, Elahi R, Esmaeilzadeh A. Clinical Applications of Interleukin-37: A Key Player in the Immunopathogenesis of Immune Disorders. Iranian journal of allergy, asthma, and immunology. 2020;19(3):209-228.

[32] Xie L, Huang Z, Li H, Liu X, Guo Zheng S, Su W. IL-38: A new player in inflammatory autoimmune disorders. Biomolecules. 2019;9(8):345.

[33] Ren X, Gelinas AD, von Carlowitz I, Janjic N, Pyle AM. Structural basis for IL-1α recognition by a modified DNA aptamer that specifically inhibits IL-1α signaling. Nature communications. 2017;8(1):1-13.

[34] Behzadi P, Ranjbar R. Caspases and Apoptosis. Molecular Enzymology and Drug Targets. 2015;1(2):1-4.

[35] Barksby H, Lea S, Preshaw P, Taylor J. The expanding family of interleukin-1 cytokines and their role in destructive inflammatory disorders. Clinical & Experimental Immunology. 2007;149(2):217-225.

[36] Gabay C, Lamacchia C, Palmer G. IL-1 pathways in inflammation and human diseases. Nature Reviews Rheumatology. 2010;6(4):232.

[37] Weber A, Wasiliew P, Kracht M. Interleukin-1 (IL-1) pathway. Science signaling. 2010;3(105):cm1-cm.

[38] Kuo W-C, Lee C-C, Chang Y-W, Pang W, Chen H-S, Hou S-C, et al. Structure-based Development of Human Interleukin-1β-Specific Antibody That Simultaneously Inhibits Binding to Both IL-1RI and IL-1RAcP. Journal of Molecular Biology. 2021;433(4):166766.

[39] Dinarello CA. Interleukin-1 in the pathogenesis and treatment of inflammatory diseases. Blood, The Journal of the American Society of Hematology. 2011;117(14):3720-3732.

[40] Elaraj DM, Weinreich DM, Varghese S, Puhlmann M, Hewitt SM, Carroll NM, et al. The role of interleukin 1 in growth and metastasis of human cancer xenografts. Clinical Cancer Research. 2006;12(4):1088-1096.

[41] Arend WP, Palmer G, Gabay C. IL-1, IL-18, and IL-33 families of cytokines. Immunological reviews. 2008;223(1):20-38.

[42] Dinarello C, Novick D, Kim S, Kaplanski G. Interleukin-18 and IL-18 binding protein. Frontiers in immunology. 2013;4:289.

[43] Fabbi M, Carbotti G, Ferrini S. Context-dependent role of IL-18 in cancer biology and counter-regulation by IL-18BP. Journal of leukocyte biology. 2015;97(4):665-675.

[44] OKAMURAH TH, KOMATSU T. Cloning of a new cytokine that induces interferon-γ. Nature. 1995;378:88-91.

[45] Pastille E, Wasmer M-H, Adamczyk A, Vu VP, Mager LF, Phuong NNT, et al. The IL-33/ST2 pathway shapes the regulatory T cell phenotype to promote intestinal cancer. Mucosal immunology. 2019;12(4):990-1003.

[46] Coyle AJ, Lloyd C, Tian J, Nguyen T, Erikkson C, Wang L, et al. Crucial role of the interleukin 1 receptor family member T1/ST2 in T helper cell type 2–mediated lung mucosal immune responses. The Journal of experimental medicine. 1999;190(7):895-902.

[47] Villacorta H, Maisel AS. Soluble ST2 testing: a promising biomarker in the management of heart failure. Arquivos brasileiros de cardiologia. 2016;106(2):145-152.

[48] Moussion C, Ortega N, Girard J-P. The IL-1-like cytokine IL-33 is constitutively expressed in the nucleus of endothelial cells and epithelial cells in vivo: a novel 'alarmin'? PloS one. 2008;3(10):e3331.

[49] Lefrançais E, Roga S, Gautier V, Gonzalez-de-Peredo A, Monsarrat B, Girard J-P, et al. IL-33 is processed into mature bioactive forms by neutrophil elastase and cathepsin G. Proceedings of the National Academy of Sciences. 2012;109(5):1673-1678.

[50] Malik A, Sharma D, Zhu Q, Karki R, Guy CS, Vogel P, et al. IL-33 regulates the IgA-microbiota axis to restrain IL-1α–dependent colitis and tumorigenesis. The Journal of clinical investigation. 2016;126(12):4469-4481.

[51] Dinarello C, Arend W, Sims J, Smith D, Blumberg H, O'Neill L, et al. IL-1 family nomenclature. Nature immunology. 2010;11(11):973-.

[52] Gresnigt MS, van de Veerdonk FL, editors. Biology of IL-36 cytokines and their role in disease. Seminars in immunology; 2013: Elsevier.

[53] Towne JE, Renshaw BR, Douangpanya J, Lipsky BP, Shen M, Gabel CA, et al. Interleukin-36 (IL-36)

ligands require processing for full agonist (IL-36α, IL-36β, and IL-36γ) or antagonist (IL-36Ra) activity. Journal of Biological Chemistry. 2011;286(49):42594-42602.

[54] Wang X, Zhao X, Feng C, Weinstein A, Xia R, Wen W, et al. IL-36γ transforms the tumor microenvironment and promotes type 1 lymphocyte-mediated antitumor immune responses. Cancer cell. 2015;28(3):296-306.

[55] Boraschi D, Lucchesi D, Hainzl S, Leitner M, Maier E, Mangelberger D, et al. IL-37: a new anti-inflammatory cytokine of the IL-1 family. European cytokine network. 2011;22(3):127-147.

[56] Taylor SL, Renshaw BR, Garka KE, Smith DE, Sims JE. Genomic organization of the interleukin-1 locus. Genomics. 2002;79(5):726-733.

[57] van de Veerdonk FL, de Graaf DM, Joosten LA, Dinarello CA. Biology of IL-38 and its role in disease. Immunological reviews. 2018;281(1):191-196.

[58] van de Veerdonk FL, Stoeckman AK, Wu G, Boeckermann AN, Azam T, Netea MG, et al. IL-38 binds to the IL-36 receptor and has biological effects on immune cells similar to IL-36 receptor antagonist. Proceedings of the National Academy of Sciences. 2012;109(8):3001-3005.

[59] Gajdács M, Dóczi I, Ábrók M, Lázár A, Burián K. Epidemiology of candiduria and Candida urinary tract infections in inpatients and outpatients: Results from a 10-year retrospective survey. Central European journal of urology. 2019;72(2):209.

[60] Leonard WJ, Lin J-X, O'Shea JJ. The γc family of cytokines: basic biology to therapeutic ramifications. Immunity. 2019;50(4):832-850.

[61] Liao W, Lin J-X, Leonard WJ. IL-2 family cytokines: new insights into the complex roles of IL-2 as a broad regulator of T helper cell differentiation. Current opinion in immunology. 2011;23(5):598-604.

[62] Leonard WJ. Cytokines and immunodeficiency diseases. Nature Reviews Immunology. 2001;1(3):200-208.

[63] Razavi GSE, Allen T. Emerging role of interleukins in cancer treatment. Immunome Research. 2015;11(S2):1.

[64] Van Dyken SJ, Locksley RM. Interleukin-4-and interleukin-13-mediated alternatively activated macrophages: roles in homeostasis and disease. Annual review of immunology. 2013;31:317-343.

[65] Kaplan MH, Hufford MM, Olson MR. The development and in vivo function of T helper 9 cells. Nature Reviews Immunology. 2015;15(5):295-307.

[66] Wang X, Lupardus P, LaPorte SL, Garcia KC. Structural biology of shared cytokine receptors. Annual review of immunology. 2009;27:29-60.

[67] Hallett MA, Venmar KT, Fingleton B. Cytokine stimulation of epithelial cancer cells: the similar and divergent functions of IL-4 and IL-13. Cancer research. 2012;72(24):6338-6343.

[68] Akdis M, Aab A, Altunbulakli C, Azkur K, Costa RA, Crameri R, et al. Interleukins (from IL-1 to IL-38), interferons, transforming growth factor β, and TNF-α: Receptors, functions, and roles in diseases. Journal of Allergy and Clinical Immunology. 2016;138(4):984-1010.

[69] Yuzhalin A, Kutikhin A. Interleukins in cancer biology: their heterogeneous role: Academic Press; 2014.

[70] Zarogoulidis P, Lampaki S, Yarmus L, Kioumis I, Pitsiou G, Katsikogiannis N, et al. Interleukin-7 and interleukin-15 for cancer. Journal of Cancer. 2014;5(9):765.

[71] Ozaki K, Spolski R, Feng CG, Qi C-F, Cheng J, Sher A, et al. A critical role for IL-21 in regulating immunoglobulin production. Science. 2002;298(5598):1630-1634.

Chapter 2

Evolutionary Conservation of the Role of CD4 as a Receptor for Interleukin-16

Gregory D. Maniero

Abstract

The interaction of CD4 with MHC class II during helper T-cell activation and effector function is required for the initiation of an adaptive immune response in all gnathostomes. CD4 is comprised of four immunoglobulin domains but most likely arose from an ancestral two-domain homolog. The distal, D1 domain of CD4 binds to non-polymorphic regions of the MHC molecule, but despite the absolute requirement for this interaction, the sequence and structure of this domain are not well conserved through phylogeny. Conversely, the proximal, D4 domain of CD4 contains the binding site of the cytokine IL-16 and is highly conserved in its amino acid structure. IL-16 is a cytokine that has been described in a wide variety of invertebrate and vertebrate species. The CD4-binding residues on IL-16 are highly conserved throughout phylogeny, allowing for promiscuous binding of IL-16 to CD4 between members of unrelated taxa. This chapter aims to present structural, and functional support for the hypothesis that the CD4 co-receptor of the TCR arose from a primordial receptor for IL-16.

Keywords: IL-16, CD4, evolution, chemoattraction, T cells

1. Introduction

The importance of acquired immunity for the survival of an organism in the face of an environment full of potential pathogens cannot be understated. The immune functions essential to acquired immunity arose with the vertebrates [1–3] and rely heavily on the activation of a $CD4^+$ subset of T lymphocytes. The role of $CD4^+$ lymphocytes is well known and seems to have appeared with gnathostomes. Although T-helper (Th) cells require interactions between CD4 and classical MHC class II, both molecules most likely did not arise at the same point in evolution. A growing body of evidence suggests that CD4 was originally a receptor for Interleukin-16 (IL-16) that was integrated into the immune system as a co-receptor of the T-cell receptor (TCR) complex. The aim of this chapter is to present some of the structural and functional characteristics that have been retained in CD4 among diverse vertebrate taxa, and the same type of characteristics retained in IL-16 throughout phylogeny, including in species much older than vertebrates.

2. Il-16

IL-16 was first described in 1982 as Lymphocyte Chemoattraction Factor (LCF) for its ability to recruit lymphocytes independent of antigen specificity to the site of inflammation during a delayed-type hypersensitivity (DTH) reaction [4–8]. IL-16 is a unique [6, 9] pleiotropic cytokine that is secreted primarily by $CD8^+$ T cells but can also be produced by eosinophils, dendritic cells (DCs), monocytes, macrophages, mast cells, as well as activated $CD4^+$ T cells, fibroblasts, and bronchial epithelial cells [10–14]. Production of IL-16 can be stimulated by mitogens or histamines as well as by recognition of antigen [4–6, 9, 15]. IL-16 influences pathological states including asthma and multiple sclerosis. Additionally, IL-16 mRNA is often present in lymphoid organs, suggesting a role in normal immune function [11].

IL-16 is post-translationally cleaved by caspase 3 from the C-terminal end of a 68-kDa pro-IL-16 molecule [16–18]. Active IL-16 is a 17 kDa protein that contains a single PDZ domain [6, 7]. Unlike in other PDZ-containing proteins, this domain is not functional for protein binding as it is partially blocked by a nearby tryptophan side-chain [19]. Native IL-16 is released as a monomer and as a tetramer, which is the primary active form of the molecule [9]. A GLGF motif within the PDZ domain may be important for the oligomerization of IL-16 [6, 17–19] and secreted monomeric IL-16 will auto-aggregate to spontaneously form active tetramers [11]. IL-16 is stored in its active configuration in granules of $CD8^+$T cells that are the main contributors of this cytokine in an immune response [16–18]. Although tetrameric IL-16 is the primary active form, monomeric IL-16 is capable of binding to CD4 and can induce a variety of phenotypic changes in target cells [9, 16].

3. IL-16 as a ligand for CD4

IL-16, whether in monomeric or tetrameric configuration, most notably attracts $CD4^+$ cells and most IL-16-mediated cell migration has been demonstrated in human and murine lymphocytes. The only described receptor for IL-16 is the CD4 co-receptor of the TCR complex, and IL-16-induced lymphotaxis is strictly limited to $CD4^+$ cells [6, 14], as evidenced by the fact that $CD4^+$, but not $CD4^-$ T cells migrate in response to IL-16 [9, 20]. Additionally, the degree of IL-16-induced chemotaxis in vitro is proportional to the density of CD4 on the surface of the responding lymphocytes [9]. The chemoattractant activity of recombinant IL-16 (rIL-16) is blocked by preincubation with F_{ab} fragments of the anti-CD4 mAb OKT4 [9, 15]. Furthermore, IL-16 elicits the migration of Th1 lymphocytes more than that of Th2 cells [19]. Recombinant human IL-16 (rhIL-16) initiates T-cell chemotaxis, in vitro, at 10 nM rhIL-16 to as low as 0.001 nM with a 50% effective dose (ED_{50}) of 0.01 nM [6, 20, 21].

Upon binding to CD4, IL-16 induces a variety of changes on Th cells in addition to chemotaxis. In addition to being a growth factor for $CD4^+$ T cells, IL-16 synergizes with IL-2 to promote the expansion of T-cell populations [12]. The binding of CD4 by IL-16 stimulates T lymphocytes to upregulate the production of various secreted and surface-bound proteins. Following incubation with IL-16, $CD4^+$ lymphocytes increase IL-2R on the cell surface [7, 9, 12], which indicates that IL-16 is involved in the expansion of T-cell populations independent of the recognition of antigen on MHC. Both native and recombinant IL-16 can induce expression of classical MHC class II molecules (primarily HLA-DR) on the membrane of non-stimulated human $CD4^+$ T cells in vitro [7, 9, 16] as well as stimulate them to produce GM-CSF [12]. Incubation with rhIL-16 will cause conA-stimulated, human $CD4^+$ MHC class II^+ PBMCs to down-regulate production of IL-2 [22]. By interfering with the interaction of CD4 with the TCR complex, rIL-16 effectively disrupts mixed lymphocyte reactions (MLR, [15]). Attraction of $CD4^+$ Th cells is induced by both native IL-16 and monomeric rIL-16 [4, 5].

4. IL-16 is an ancient and ubiquitous cytokine

IL-16 is a cytokine that is found ubiquitously throughout vertebrate phylogeny and has been described not only in humans but in a variety of mammals [20], birds [23–26], fish [27, 28]. IL-16 has even been described in invertebrates [29–31] Additionally, inferred protein sequences from genetic data can be found for IL-16 and pro-IL16 in amphibians and reptiles [32]. The sequence and structure of IL-16 and pro-IL-16 show considerable homology among disparate vertebrate groups [13, 32, 33]. In mammals, IL-16 is highly conserved among humans, mice, and African green monkeys at the structural and genetic levels [6, 16]. Conserved residues on IL-16 that have been determined to be important for binding to CD4 by competitive binding assays are arginines at positions equivalent to human 106 and 107, and possibly a leucine at the equivalent of position 108 [13, 19].

5. CD4 as a receptor for IL-16

Although this chapter focuses on its functions as receptor for IL-16, CD4 is predominantly known for its role as a co-receptor in the T-cell receptor complex. As a co-receptor, CD4 binds, in conjunction with the α:β TCR, to MHC class II during antigen-dependent helper T-cell (Th) activation, differentiation, and effector function to substantially increase TCR-signaling [34, 35]. CD4 is a non-polymorphic 55-kDa monomer that consists of four extramembrane, Ig-like regions named from the distal D1 to the proximal D4 region [17, 36–39]. The D1 and D3 regions closely resemble V-type immunoglobulin domains, and D2 and D4 resemble C-type domains [34, 38, 40]. The CD4 co-receptor is expressed on the surface of Th cells and on some subsets of neutrophils and monocytes. Several models have been proposed to explain the association between CD4 and the α:β TCR on Th cells however the functional result of all of them is that CD4 associates with the TCR and enhances effector function [41]. The distal D1 region of CD4 interacts with the non-polymorphic region on the β2 domain of MHC class II molecules [39, 42]. This CD4:MHC association appears to stabilize and increase the affinity of TCR binding to antigen-bearing MHC class II molecules expressed on the surface of antigen presenting cells (APC) [42–44]. Additionally, CD4 augments intracellular signaling by recruiting the intracellular kinase $p56^{lck}$ that enhances the phosphorylation of ITAMs (immunoreceptor tyrosine-based activation motifs) upon the engagement of a TCR with its cognate antigen:MHC molecular complex. The phosphorylated ITAMs recruit ZAP70 that, when phosphorylated by $p56^{lck}$, continues downstream T-cell activation events [45]. The inclusion of CD4 in the immune synapse is necessary for effective T-cell effector function mediated by signaling through the TCR complex [34, 37].

A typical Th response begins with the engagement of the TCR with its cognate antigen:MHC complex. Complete cellular activation requires interaction with the CD4 co-receptor, the CD3 tetramer, and the intracellular ζ-chain. Binding of CD4 without subsequent co-receptor signaling can result in incomplete T-cell activation, limited IL-2-mediated proliferation, and eventual anergy. Crosslinking of the CD4 co-receptor results in downstream signaling that is independent of antigen, the TCR, and CD3. Partial activation can occur with cross-linking of CD4 by anti-CD4

$F(ab')_2$ fragments in the absence of antigen recognition or cell-to-cell contact. As in normal $CD4^+$ T-cell signaling, the low-level stimulation of exclusively CD4 engagement is due to the recruitment and activation of $p56^{lck}$ [10]. Phosphorylation and activation of $p56^{lck}$ initiates cell migration and increases in intracellular NFκB, IP_3, and Ca^{2+} as well as nuclear translocation of PKC [6, 10, 12, 15]. Treating $CD4^+$ immune cells with IL-16 results in a cell phenotype that bears a striking resemblance to that seen following anti-CD4 treatment [6].

Cross-linking of CD4 by tetrameric IL-16 or binding by monomeric protein results in the association of CD4 with, and the phosphorylation of $p56^{lck}$ [9, 10]. Transfection with human CD4 allows murine hybridomas to migrate in response to IL-16, but this response is absent in cells transfected with a mutant CD4 variant that is unable to associate with cytoplasmic $p56^{lck}$ [10, 20]. Native, but not recombinant, IL-16 stimulates $CD4^+$ T cell proliferation [7, 18], whereas rhIL-16 stimulates $CD4^+$ lymphocyte progression from G_0 to G_{1a} but does not initiate proliferation [6, 16, 22]. Both native and recombinant IL-16 will induce the expression of MHC class II (specifically HLA-DR) on the surface of resting human $CD4^+$ lymphocytes [7, 9, 42] and can induce their production of GM-CSF in vitro [12]. Since IL-16 can spontaneously form tetramers following release, it is difficult to tease out the difference between activation by monomeric and tetrameric IL-16. In addition to initiating signaling through CD4, IL-16 blocks the interaction of the CD4 co-receptor with the TCR complex. In fact, this is the mechanism that is responsible for IL-16-mediated disruption of in vitro MLR [15].

6. IL-16 preferentially recruits and activates regulatory T cells

In part due to its ability to induce lymphocyte migration, IL-16 is classified as a pro-inflammatory cytokine, yet it appears to slow TCR-mediated activation [7]. As a chemoattractant of $CD4^+$ lymphocytes, IL-16 appears to preferentially attract and activate regulatory T cells (Tregs), which suppress T cell activity [21]. IL-16 inhibits the production of IL-2 by mitogen-activated $CD4^+$ lymphocytes in humans and preferentially attracts lymphocytes that express mRNA for FoxP3 in vitro [22]. During inflammatory lung injury, IL-16 produced in part by the lung endothelium, attracts $CD4^+$ T cells that express FoxP3, produce IL-10, and act to protect the lungs from infiltration by neutrophils [46]. T cells that migrate in vitro in response to IL-16 in transwell experiments express more CD25 and CTLA-4 on their surface and release more TGFβ than control cells. In addition, cells that migrate in response to IL-16 express higher levels of FoxP3 mRNA and protein than do control cells [21], suggesting that IL-16 primarily attracts T cells with a regulatory phenotype. Recombinant rhIL-16, as well as recombinant *Xenopus* IL-16 (rXIL-16, Maniero, unpublished data), recruits lymphocytes to the body cavity of *Xenopus laevis*. Upon examination, the recovered lymphocytes are seen to express mRNA for CD4 to a greater extent than for CD8, CTLA-4 more than CD28, as well as both FoxP3 and IL-10, suggesting a regulatory phenotype for the IL-16-recruited lymphocytes [47]. These mRNA levels were found in cells that were recovered approximately 16 h post injection with IL-16, so it is impossible to distinguish, from these experiments, if regulatory cells are attracted by IL-16 or if IL-16 induces the expression of a suite of regulatory genes [47].

7. Conservation of CD4

CD4 is highly conserved in mammals, yet the primary and secondary structures vary considerably among vertebrates [32, 38]. In the distal, D1 region, the canonical MHC class II-binding motif of FLXK is found on all eutherian mammals that have been studied [32, 38]. However, even though all gnathostomes demonstrate CD4-dependent Th-activity, the D1 region is not highly conserved among representatives of disparate vertebrate groups [32, 38]. Most non-mammalian vertebrates do not possess the classic FLXK MHC class II-binding motif seen in mammals [32] yet rely heavily on traditional T-helper activity, suggesting that the role of CD4 binding to classical MHC molecules is an important function that has arisen on multiple occasions in vertebrate evolution. Other conserved motifs on the CD4 molecule are not involved with MHC class II binding. These conserved regions include a WXC motif and the intracellular CxC motif that associates with $p56^{lck}$ in the cytosol [34, 38]. The association of $p56^{lck}$ with CD4 is imperative for Th cell activation and the conservation of the lck-binding site demonstrates the essential and primordial association between these molecules [41, 48].

Although CD4 binds to ligands other than MHC class II molecules, including the surface glycoprotein gp120 of the Human Immunodeficiency Virus (HIV), which binds outside of the MHC class II-binding domain [34, 42, 43], the only described physiological role for the proximal, D4 region of CD4 is that of a receptor for the cytokine IL-16 [6, 19]. The IL-16 binding site on the proximal D4 domain of CD4 is an effectively long distance from the MHC-binding site [6, 19, 34]. On human CD4, there are two points on the D4 domain that are important for the binding of IL-16. A WQCLL motif from amino acids 344–348 is of major importance, but two valines at position 334 and 336 have been shown to be important in humans [33].

Amino acid sequence alignments, produced with CLUSTALW (www.genome.jp), show that the proximal D4domain of CD4, and especially the binding site for IL-16, is highly conserved (**Figure 1**), allowing for promiscuous binding of IL-16 to CD4 between disparate gnathostomes. In fact, human IL-16 recruits

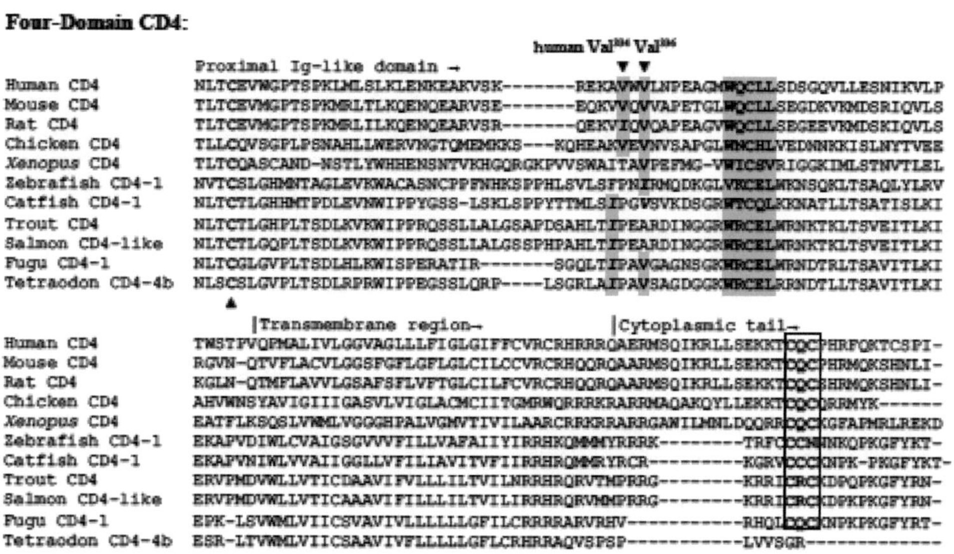

Figure 1.
*Multiple alignment (CLUSTALW) of deduced amino acid partial sequences from CD4 D4 region of several vertebrates. The top portion presents the proximal, D4 domain of CD4. Conserved cysteines near the N-terminus of the proximal Ig domain are bolded and marked with an ▲. The canonical mouse and human WQCLLS motif that corresponds to human Val334, Val336, Gln345, is shaded, as are conserved residues that occupy the positions of and Leu347 and Leu347 required for binding IL-16 on human CD4 [19]. The bottom section shows amino acids from the transmembrane region and the cytoplasmic tail of CD4. The box surrounds the putative (CxC) binding site for p56lck . (human:*Homo sapiens *NCBI accession no. NP_000607.1, mouse:* Mus musculus *NP_038516.1, rat:* Rattus novegicus *NP_036837.1, chicken* Gallus gallus *CAA72740.1,* Xenopus laevis *NP_001233240.1, zebrafish CD4–1:* Danio rerio *XP_005173553.1, catfish CD4–1:* Ictalurus punctatus *NP_001187155.1, trout CD4:* Oncorhynchus mykiss *AAY42070.1, Salmon CD4-like* Salmo salar *XP_014019051.1, Fugo CD4–1;* Takifugu rubripes *NP_001072091.1, Tetraodon CD4–4b;* Tetraodon nigroviridis *ABU95654.1.*

murine CD4⁺ lymphocytes in vitro, and mouse IL-16 similarly recruits human CD4⁺ lymphocytes [16]. In mice, IL-16-induced chemotaxis of CD4⁺ lymphocytes is blocked by the addition of anti-human IL-16 antibodies [16]. As would be expected with the conservation of the IL-16 binding region on CD4, IL-16 from derived vertebrates can activate CD4⁺ T lymphocytes from more ancestral organisms [32]. Recombinant human IL-16 (rhIL-16) binds to lymphocytes from the African Clawed Frog, *Xenopus laevis* with sufficient avidity to allow rhIL16-bound lymphocytes to be separated on magnetic columns [32]. No reagents exist that can positively identify CD4 on *Xenopus* cells [3], and magnetic bead separation can merely suggest that rhIL-16 is binding to *Xenopus* CD4. Monoclonal antibodies specific for *Xenopus* CD8, however are available and can be used to isolate *Xenopus* CD8⁺ T cells [3]. Incubation of *Xenopus* lymphocytes with rhIL-16 in vitro, correlates with the expression of MHC class II mRNA by CD8⁻ cells but not those that are CD8⁺, indicating that rhIL-16 is most likely binding to *Xenopus* CD4⁺ lymphocytes [32]. As explained earlier, the role originally attributed to IL-16 was lymphocyte attraction, and injection of rhIL-16 into the body cavity of the amphibian *Xenopus* leads to the accumulation of lymphocytes in the peritoneum [32, 47]. The cells that are recruited to the *Xenopus* body cavity by rhIL-16 express mRNA for CD4 to a greater extent than that for CD8α or CD8β [32], again suggesting that rhIL-16 is recruiting CD4⁺ *Xenopus* lymphocytes. The ability of IL-16 to affect CD4 cells from members of disparate vertebrate groups is hardly surprising. Not only is the IL-16-binding site highly conserved on CD4 (**Figure 1**), but the region of IL-16 that binds to CD4 is highly conserved throughout phylogeny (**Figure 2**).

```
Human            SASAASDVSVESTAEATVCTVTLEKMSAGLGFSLEGGKGSLHGDKPLTINRIFKGAASE
Mouse            SASAASDISVE-SKEATACTVTLEKTSAGLGFSLEGGKGSLHGDKPLTINRIFKGTE--
Rat              SASVASGISVE-SVEATVCTVTLEKTSAGLGFSLEGGKGSLHGDKPLTINRIFKGTE--
Chicken          TSSVASDASQESTTEETICTITLDKTAAGLGFSLEGGKGSIHGDKPIIINRIFKGTSLE
Xenopus laevis   SALTDDSTAANCDDTGDIIVVTLEKSLAGLGFSLDGGKGSVQGDRPVIINRIFKGVS-E
Rainbow Trout    VSSSSTDLNPTVEDGGIMLTLELETGGGGVGFSLDGGKGSIHGDRPLVINRIFKGGAAE
Puffer fish      --------GVTVEVQGGPITVKIIKGAAGVGFTLEGGKGSIHGDRPLVINRIFTDDD--
Shrimp           SKSLVSDTESGLVPRGPPLTITLVKDGAGLGFSLEGGKDSPLGDRPLTVKKVFSGGAAD
Mitten crab      SKSLISDVAAMSVPRGPPFTITLVKDGAGLGFSLEGGKDSPLGDRPLTVKKIFSGGAAD
Mud Crab         SKSIISDVAAMSVPRGPPFTITLVKDGAGLGFSLEGGKDSPLGDRPLTVKKIFSGGAAD
Leech            ------------------TVKLEKGFLGVGFCIEGGRASPYGDKPILIKRVVRGDVP-

Human            QSETVQPGDEILQLGGTAMQGLTRFEAWNIIKALPDGPVTIVIRRKSLQSKETTAAGDS
Mouse            QGEMVQPGDEILQLAGTAVQGLTRFEAWNVIKALPDGPVTIVIRRTSLQCKQTTASADS
Rat              QGDAVQPGDEILQLAGTAVQGLTRFEAWNVIKALPDGPVTVVIRRNGLEGKQTTASADL
Chicken          QSSPVQPGDELLQVHTTALQGLTRFEAWNIIKALPDGPITAIIRRKNPSSVTKKASETL
Xenopus laevis   KNNAVQYGDELLQLGNISLQGLTRFEAWNSIKSLPNGLVQAVIRRKSTDSTVNQ-----
Rainbow Trout    QSGLQS-GDELLQVQSTSLQELSRFEAWNIVKALPEGHITLVIRRRKEDEAEGSA----
Puffer fish      ---ALKMGDVLLQVQDVSVQEMTRFEAWNLVKSLPEGPVTVVIARKTGAAE--------
Shrimp           KGGILKVGDELVSVNTVDVTSMARIEAWNFLKKLPDGTVSLVLRQKVEETQMKSEE---
Mitten crab      KGGVLKVGDELVSVNTVDVTGMARIEAWNFLKKLPDGTVTLVLRQKLDYTAAKNEE---
Mud Crab         KGGILKVGDELVSVNAVDVTGMARIEAWNFLKKLPDGTVTLVLRQKLDGS---------
Leech            ----LRAGDELVSVNGKEVSSMTSAQVWSLLKSLPGGVMVLELRRKE------------
```

Figure 2.

*Multiple alignment (CLUSTALW) of IL-16 deduced amino acids from several different vertebrates. The conserved GLGF binding cleft of the PDZ domain highlighted in gray. The residues that are critical for binding to domain 4 of CD4 to initiate chemotaxis, Argenines106–107 and Lysine108, are conserved throughout phylogeny and are underlined and bolded. (human:*Homo sapiens *NCBI accession no. AAC12732.1, mouse:* Mus musculus *AAC16039.1, rat:* Rattus novegicus *XP_006229550.1, chicken:* Gallus gallus *NP_001264925.1,* Xenopus laevis *XP_018108634.1, rainbow trout:* Oncorhynchus mykiss *CAD70074.2, puffer fish:* Tetraodon nigroviridis *AAX36076.3, shrimp:* Penaeus vannamei *ASJ26360.1, mitten crab:* Eriocheir sinensis, *Mud crab: (Gu et al., 2017), leech:* Hirudo medicinalis *ACF07997.1.*

8. The binding domains for IL-16:CD4 interactions are highly conserved

IL-16 is a cytokine that is produced by many organisms. Two residues crucial for binding to CD4 and homologous to human arg$^{106-107}$, are highly conserved in IL-16 from mammals, birds, amphibians, and teleost fish (**Figure 2**) [13, 19, 32]. Additionally, the GLF cleft of the IL-16 PDZ domain necessary for IL-16 oligomerization (**Figure 2**) [6, 16, 18, 19] is highly conserved throughout phylogeny [32]. The conservation of the D4 region of vertebrate CD4, along with the conservation of vertebrate IL-16, especially the conserved arginine residues that bind to CD4, certainly explains the intraspecies promiscuity of the IL-16:CD4 binding and activation [32].

In addition to vertebrates, IL-16 or proIL-16 has been described in several species of invertebrates, including the Chinese Mitten Crab *Eriocheir sinensis,* [49]), the mud crab (*Scylla paramosain*, [29]), and the Pacific white shrimp (*Litopenaeus vannamei*, [30]). A homolog of IL-16 has even been described from the nervous system of the European medicinal leech, *Hirudo medicinalis* [31]. The amino acid sequences of invertebrate IL-16 homologs indicate that these molecules contain the conserved arginine residues necessary for interaction with CD4 (**Figure 2**). These residues are equidistant from the PDZ domain in vertebrate IL-16 in all of the sequences that we examined. This is of particular importance since these organisms lack CD4, a molecule that is only found in vertebrates. The conservation of these residues on ancestral organisms has not been demonstrated previously, but strongly argues for the existence of a receptor for IL-16 that has at least some similarity to the D4 domain of vertebrate CD4.

Jawed vertebrates possess similar adaptive immune systems that rely on helper T-cell effector functions that depend, in a large part, upon CD4-MHC class II interactions. A vast majority of jawed vertebrates, with some notable exceptions, express CD4 on a subset of lymphocytes despite the fact that genes for CD4 are not well conserved among disparate species, even if they are closely related [50]. The gene for CD4 has been described and cloned in many species of teleost fish [51–58]. Helper, CD4$^+$ T cells in teleost function in a manner similar, if not identical to that found in their tetrapod counterparts. Like those in tetraposds, teleost T cells develop in the thymus as CD4$^+$CD8$^+$ double-positive cells migrate to the periphery as single positive lymphocytes. As in mammals, teleost CD4$^+$ T cells proliferate in mixed lymphocyte reactions and in an antigen-specific manner [55]. Helper T-cell function has been documented in fish, and adoptive transfer of CD4$^+$ cells from sensitized fish enhances virus-specific antibody formation [59].

9. Two-domain, CD4-like molecules

Two discrete forms of CD4 have been described in teleosts; one, consisting of four immunoglobulin-like domains that folds in a manner similar to that of tetrapod CD4 and interacts with classical MHC class II molecules [57], and a second type of CD4 molecule, that consists of only two immunoglobulin-like domains that does not appear to possess the ability to interact with MHC class II that is referred to as CD4–2, CD4REL, or CD4-like [50]. These two-domain CD4 molecules are expressed on the surface of a limited subset of teleost T cells and have cytoplasmic tails that associate with kinase p56^{lck} like those of canonical, four-domain CD4 molecules [40, 50, 53, 58]. A genes for a four-domain CD4 molecule has been described in lamprey, but this lamprey CD4-like molecule does no include a canonical CXC motif that is required for the interaction with p56^{lck} [60, 61]. Elasmobranchs lack genes for either two- or four-domain CD4 molecules but possess genes for both MHC class I and MHC class IIα and β. Elasmobranchs exhibit only T-cell responses of a Th1 phenotype. Additionally, these primitive cartilaginous vertebrates possess CD4/LAG3-like genes that may encode an as-yet un-described molecule that is functionally homologous to CD4 in derived vertebrates [62, 63].

The two-domain, CD4-like molecules in fish have a proximal domain (D2) that possesses structural similarity to immunoglobulin constant regions (C-like) and a distal domain (D1) more similar to Ig variable domains (V-like) [61]. As stated above, the CD4 molecule of all gnathasomes, including teleosts, consist of four immunoglobulin-like domains. Due to the repeating domain structure of this molecule it has been postulated that the genes for typical CD4 molecules are derived from a duplication of an ancestral gene that encoded a two-Ig-domain, CD4-like, cell-surface molecule [64], although this does not explain the sole, four-domain CD4-homolog seen in the ancestral cyclostomes. In *Tetraodon,* lymphocytes that express CD4–2 appear to bind and migrate in response to recombinant IL-16 preferentially compared to those that express CD4–4 [56]. Additionally, the CD4–2$^+$ *Tetraodon* lymphocytes appear to have a regulatory phenotype, expressing mRNA for FoxP3 and having a CD25-like molecule on their surface [56]. The affinity of the likely ancestral CD4–2 for IL-16, and the possibility that IL-16 recruits CD4–2$^+$ regulatory lymphocytes supports our hypothesis that four-domain CD4 arose from an ancestral, two-domain interleukin receptor.

Many, and perhaps all four-domain CD4 molecules possess amino sequences that are homologous to human to IL-16-binding residues (**Figure 1**), and it seems that similar motifs are present on the proximal domains of two-domain CD4-related proteins. Although not identical to those seen on four-domain proteins, the deduced amino acid sequences of two-domain CD4 homologs from teleosts show potential IL-16-binding motifs that are spaced equidistant from the conserved cysteine at the N-terminal region of the most proximal Ig-like domain seen in traditional CD4 molecules (**Figure 3**). In all of the teleost two-domain CD4s that we have compiled, there is a valine at a position similar to the mammalian val^{336}. Additionally, two-domain CD4 homologs possess a sequence similar to the four-domain WQCLL motif, again, in a similar, if not identical, position in the proximal Ig-like domain. Rather than WQCLL, the two-domain motif is WTCQ(or L, K, or T)I (or V, F, or P). Although not identical, these motifs in both types of CD4 have some distinct similarities. Both of the five-amino acid domains have a cysteine at the center that is found in sequences of all of the species that we examined. The first amino acid in almost all of these motifs is a tryptophan (W), and the fourth and fifth amino acids are almost always aliphatic. The second of the five usually contains an acidic residue but some variation is seen. Regardless of the differences, there is evidence of a possible IL-16-binding motif on both CD4 and CD4–2 molecules. It is interesting to note that, although the lamprey CD4-like molecule has four Ig-like domains, the proximal domain is more similar, including at the putative IL-16-binding site, to teleost two-domain molecules that to conventional CD4 (**Figure 3**).

As previously stated, the lamprey CD4-like molecule does not possess a canonical motif that associates with p56^{lck}. Unlike CD4, many cytokine receptors lack a domain for tyrosine kinases. Like two-domain CD4-like molecules, and perhaps the ancestral form of CD4, many cytokine receptors are composed of two immunoglobulin-like extracellular domains and exist as single chains on the surface of cells but, upon encountering an appropriate ligand, form dimers that initiate downstream signaling [65, 66]. The proximal immunoglobulin-like domain is essential to the dimerization of hematopoietic cytokine receptors and involves a motif of four conserved amino acids that resides towards the c-terminal end of the extracellular portion of the molecule and consists of two pairs of conserved

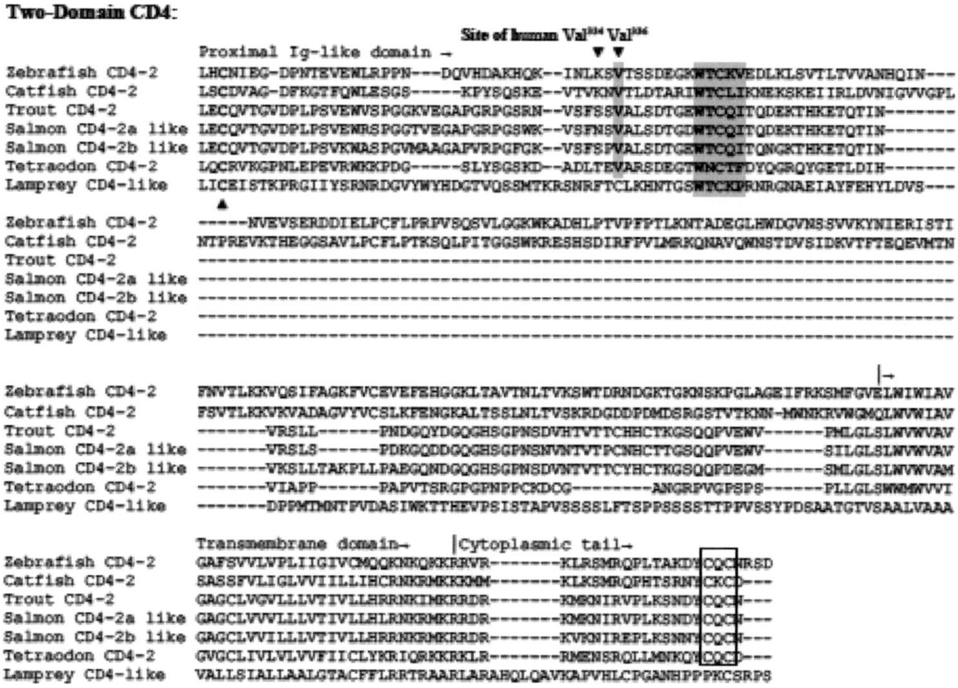

Figure 3.
*Multiple alignment (CLUSTALW) of deduced amino acid sequences from proximal domains of CD4–2 from several fish species. Conserved cysteines near the N-terminus of the proximal Ig domain are bolded and marked with an ▲. A motif similar to IL-16 binding motif of four domain CD4, is shaded, as are conserved leucines that occupy positions similar to the Leu347 required for binding IL-16 on human CD4 [19]. The box surrounds the putative (CxC) binding site for p56lck . Zebrafish CD4–2;*Danio rerio *NCBI accession no. NP_001352990.1, catfish CD4–2:* Ictalurus punctatus *NP_001187156.1, trout:* Oncorhynchus mykiss *XP_021437193.2, Salmon CD4–2a like* Salmo salar *ABZ81914.1, Salmon CD4–2b like* Salmo salar *ABZ81915.1, Tetraodon CD4–2;* Tetraodon nigroviridis *ABU95652.1, lamprey CD4-like* Petromyzon marinus *AAU09669.1.*

amino acids separated by a single, non-conserved residue (WSxWS, [65, 66]). Like these cytokine receptors, CD4 appears to form homodimers by association at the D4 domain [37, 67], the domain that contains the IL-16 binding site. Like cytokine receptors, dimerization of CD4 appears to be a prerequisite to Th activation [67]. Dimerization is also important for effective binding to IL-16, and IL-16 appears to function by bringing CD4 molecules into close proximity [6, 33, 67].

10. Conclusion: probable origins of Gnathostome CD4

Genetic and structural similarities between the D1 and D3 domains and between the D2 and D4 domains give credence to the hypothesis that vertebrate CD4 arose from a precursor with two extracellular immunoglobulin-like domains [36, 61, 64, 68]. In both agnathans and teleost fish, CD4-like molecules with two immunoglobulin domains do not associate with MHC class II molecules but nonetheless appear to be important in immune protection. Lymphocytes with two-domain CD4 molecules, such as those found in teleosts, could very well represent an ancestral CD4^{+} subset of T cells [61]. The hypothesis that CD4 arose from the duplication of a gene for a protein consisting of two extracellular immunoglobulin domains has been thoroughly discussed and supported [64]. It is quite possible that the physiological importance of these truncated CD4 molecules is that of receptors for IL-16. Regulatory T lymphocytes play a critical role in controlling the immune response. The gene for IL-16 arose well before the advent of jawed vertebrates and CD4. Although the role of IL-16 in invertebrates has not been clearly elucidated, the ancestral role for CD4 and its evolutionary precursors may be as a receptor for IL-16 that functions to regulate immune function.

Acknowledgements

The author would like to thank Acadia Kopec and Zak Michaud for critical review of this chapter, the family of Fr. Francis Hurley C.S.C. for their generous support, and Biology Department of Stonehill College.

Author details

Gregory D. Maniero
Stonehill College, Easton, MA, USA

*Address all correspondence to: gmaniero@stonehill.edu

References

[1] Flajnik MF. A cold-blooded view of adaptive immunity. Nat Rev Immunol. 2018 Jul;18(7):438-453. doi: 10.1038/s41577-018-0003-9. PMID: 29556016; PMCID: PMC6084782.

[2] Kasahara M, Suzuki T, Pasquier LD. On the origins of the adaptive immune system: novel insights from invertebrates and cold-blooded vertebrates. Trends Immunol. 2004 Feb;25(2):105-11. doi: 10.1016/j.it.2003.11.005. PMID: 15102370.

[3] Robert J, Ohta Y. Comparative and developmental study of the immune system in Xenopus. Dev Dyn. 2009 Jun;238(6):1249-70. doi: 10.1002/dvdy.21891. PMID: 19253402; PMCID: PMC2892269.

[4] Center DM, Cruikshank WW. Modulation of lymphocyte migration by human lymphokines. I. Identification and characterization of chemoattractant activity for lymphocytes from mitogen-stimulated mononuclear cells. J Immunol. 1982 Jun;128(6):2563-8. PMID: 7042840.

[5] Cruikshank W, Center DM. Modulation of lymphocyte migration by human lymphokines. II. Purification of a lymphotactic factor (LCF). J Immunol. 1982 Jun;128(6):2569-74. PMID: 7042841.

[6] Center DM, Kornfeld H, Cruikshank WW. Interleukin 16 and its function as a CD4 ligand. Immunol Today. 1996 Oct;17(10):476-81. doi: 10.1016/0167-5699(96)10052-i. PMID: 8908813.

[7] Cruikshank WW, Kornfeld H, Center DM. Signaling and functional properties of interleukin-16. Int Rev Immunol. 1998;16(5-6):523-40. doi: 10.3109/08830189809043007. PMID: 9646175.

[8] Yoshimoto T, Wang CR, Yoneto T, Matsuzawa A, Cruikshank WW, Nariuchi H. Role of IL-16 in delayed-type hypersensitivity reaction. Blood. 2000 May 1;95(9):2869-74. PMID: 10779433.

[9] Cruikshank WW, Center DM, Nisar N, Wu M, Natke B, Theodore AC, Kornfeld H. Molecular and functional analysis of a lymphocyte chemoattractant factor: association of biologic function with CD4 expression. Proc Natl Acad Sci U S A. 1994 May 24;91(11):5109-13. doi: 10.1073/pnas.91.11.5109. PMID: 7910967; PMCID: PMC43941.

[10] Ryan TC, Cruikshank WW, Kornfeld H, Collins TL, Center DM. The CD4-associated tyrosine kinase p56lck is required for lymphocyte chemoattractant factor-induced T lymphocyte migration. J Biol Chem. 1995 Jul 21;270(29):17081-6. doi: 10.1074/jbc.270.29.17081. PMID: 7615501.

[11] Chupp GL, Wright EA, Wu D, Vallen-Mashikian M, Cruikshank WW, Center DM, Kornfeld H, Berman JS. Tissue and T cell distribution of precursor and mature IL-16. J Immunol. 1998 Sep 15;161(6):3114-9. PMID: 9743378.

[12] Hermann E, Darcissac E, Idziorek T, Capron A, Bahr GM. Recombinant interleukin-16 selectively modulates surface receptor expression and cytokine release in macrophages and dendritic cells. Immunology. 1999 Jun;97(2):241-8. doi: 10.1046/j.1365-2567.1999.00786.x. PMID: 10447738; PMCID: PMC2326843.

[13] Richmond J, Tuzova M, Cruikshank W, Center D. Regulation of cellular processes by interleukin-16 in homeostasis and cancer. J Cell Physiol. 2014 Feb;229(2):139-47. doi: 10.1002/jcp.24441. PMID: 23893766.

[14] Cruikshank W, Little F. Interleukin-16: the ins and outs of regulating T-cell activation. Crit Rev Immunol. 2008;28(6):467-83. doi: 10.1615/critrevimmunol.v28.i6.10. PMID: 19265505.

[15] Theodore AC, Center DM, Nicoll J, Fine G, Kornfeld H, Cruikshank WW. CD4 ligand IL-16 inhibits the mixed lymphocyte reaction. J Immunol. 1996 Sep 1;157(5):1958-64. PMID: 8757315.

[16] Keane J, Nicoll J, Kim S, Wu DM, Cruikshank WW, Brazer W, Natke B, Zhang Y, Center DM, Kornfeld H. Conservation of structure and function between human and murine IL-16. J Immunol. 1998 Jun 15;160(12):5945-54. PMID: 9637508.

[17] Wu DM, Zhang Y, Parada NA, Kornfeld H, Nicoll J, Center DM, Cruikshank WW. Processing and release of IL-16 from CD4+ but not CD8+ T cells is activation dependent. J Immunol. 1999 Feb 1;162(3):1287-93. PMID: 9973381.

[18] Center DM, Kornfeld H, Ryan TC, Cruikshank WW. Interleukin 16: implications for CD4 functions and HIV-1 progression. Immunol Today. 2000 Jun;21(6):273-80. doi: 10.1016/s0167-5699(00)01629-7. PMID: 10825739.

[19] Nicoll J, Cruikshank WW, Brazer W, Liu Y, Center DM, Kornfeld H. Identification of domains in IL-16 critical for biological activity. J Immunol. 1999 Aug 15;163(4):1827-32. PMID: 10438915.

[20] Lynch EA, Heijens CA, Horst NF, Center DM, Cruikshank WW. Cutting edge: IL-16/CD4 preferentially induces Th1 cell migration: requirement of CCR5. J Immunol. 2003 Nov 15;171(10):4965-8. doi: 10.4049/jimmunol.171.10.4965. PMID: 14607889.

[21] McFadden C, Morgan R, Rahangdale S, Green D, Yamasaki H, Center D, Cruikshank W. Preferential migration of T regulatory cells induced by IL-16. J Immunol. 2007 Nov 15;179(10):6439-45. doi: 10.4049/jimmunol.179.10.6439. PMID: 17982032.

[22] Ogasawara H, Takeda-Hirokawa N, Sekigawa I, Hashimoto H, Kaneko Y, Hirose S. Inhibitory effect of interleukin-16 on interleukin-2 production by CD4+ T cells. Immunology. 1999 Feb;96(2):215-9. doi: 10.1046/j.1365-2567.1999.00693.x. PMID: 10233698; PMCID: PMC2326730.

[23] Min W, Lillehoj HS. Identification and characterization of chicken interleukin-16 cDNA. Dev Comp Immunol. 2004 Feb;28(2):153-62. doi: 10.1016/s0145-305x(03)00133-2. PMID: 12969800.

[24] Hong YH, Lillehoj HS, Lee SH, Dalloul RA, Lillehoj EP. Analysis of chicken cytokine and chemokine gene expression following Eimeria acervulina and Eimeria tenella infections. Vet Immunol Immunopathol. 2006 Dec 15;114(3-4):209-23. doi: 10.1016/j.vetimm.2006.07.007. Epub 2006 Sep 20. PMID: 16996141.

[25] Liu WQ, Tian MX, Wang YP, Zhao Y, Zou NL, Zhao FF, Cao SJ, Wen XT, Liu P, Huang Y. The different expression of immune-related cytokine genes in response to velogenic and lentogenic Newcastle disease viruses infection in chicken peripheral blood. Mol Biol Rep. 2012 Apr;39(4):3611-8. doi: 10.1007/s11033-011-1135-1. Epub 2011 Jul 5. PMID: 21728003.

[26] Barua A, Yellapa A, Bahr JM, Adur MK, Utterback CW, Bitterman P, Basu S, Sharma S, Abramowicz JS. Interleukin 16- (IL-16-) Targeted Ultrasound Imaging Agent Improves Detection of Ovarian Tumors in Laying Hens, a Preclinical Model of Spontaneous Ovarian Cancer.

Biomed Res Int. 2015;2015:567459. doi: 10.1155/2015/567459. Epub 2015 Jun 16. PMID: 26161406; PMCID: PMC4486604.

[27] Wang P, Lu YQ, Wen Y, Yu DY, Ge L, Dong WR, Xiang LX, Shao JZ. IL-16 induces intestinal inflammation via PepT1 upregulation in a pufferfish model: new insights into the molecular mechanism of inflammatory bowel disease. J Immunol. 2013 Aug 1;191(3):1413-27. doi: 10.4049/jimmunol.1202598. Epub 2013 Jul 1. PMID: 23817423.

[28] Wang L, Jiang L, Wu C, Lou B. Molecular characterization and expression analysis of large yellow croaker (*Larimichthys crocea*) interleukin-12A, 16 and 34 after poly I:C and Vibrio anguillarum challenge. Fish Shellfish Immunol. 2018 Mar;74:84-93. doi: 10.1016/j.fsi.2017.12.041. Epub 2017 Dec 29. PMID: 29292198.

[29] Gu WB, Zhou YL, Tu DD, Zhou ZK, Zhu QH, Chen YY, Shu MA. Identification and characterization of pro-interleukin-16 from mud crab Scylla paramamosain: The first evidence of proinflammatory cytokine in crab species. Fish Shellfish Immunol. 2017 Nov;70:701-709. doi: 10.1016/j.fsi.2017.09.057. Epub 2017 Sep 23. PMID: 28951219.

[30] Liang Q, Zheng J, Zuo H, Li C, Niu S, Yang L, Yan M, Weng SP, He J, Xu X. Identification and characterization of an interleukin-16-like gene from pacific white shrimp *Litopenaeus vannamei*. Dev Comp Immunol. 2017 Sep;74:49-59. doi: 10.1016/j.dci.2017.04.011. Epub 2017 Apr 17. PMID: 28428061.

[31] Croq F, Vizioli J, Tuzova M, Tahtouh M, Sautiere PE, Van Camp C, Salzet M, Cruikshank WW, Pestel J, Lefebvre C. A homologous form of human interleukin 16 is implicated in microglia recruitment following nervous system injury in leech *Hirudo medicinalis*. Glia. 2010 Nov 1;58(14):1649-62. doi: 10.1002/glia.21036. PMID: 20578037.

[32] Gillis J, Uccello TP, Magri Z, Morris N, Maniero GD. Preliminary indications that recombinant human IL-16 attracts and stimulates lymphocytes of the amphibian, *Xenopus laevis* implying an ancestral role for CD4 as a cytokine receptor. Cytokine. 2020 Dec;136:155254. doi: 10.1016/j.cyto.2020.155254. Epub 2020 Aug 21. PMID: 32836028.

[33] Liu Y, Cruikshank WW, O'Loughlin T, O'Reilly P, Center DM, Kornfeld H. Identification of a CD4 domain required for interleukin-16 binding and lymphocyte activation. J Biol Chem. 1999 Aug 13;274(33):23387-95. doi: 10.1074/jbc.274.33.23387. PMID: 10438516.

[34] Fleury S, Lamarre D, Meloche S, Ryu SE, Cantin C, Hendrickson WA, Sekaly RP. Mutational analysis of the interaction between CD4 and class II MHC: class II antigens contact CD4 on a surface opposite the gp120-binding site. Cell. 1991 Sep 6;66(5):1037-49. doi: 10.1016/0092-8674(91)90447-7. PMID: 1889086.

[35] Yin Y, Wang XX, Mariuzza RA. Crystal structure of a complete ternary complex of T-cell receptor, peptide-MHC, and CD4. Proc Natl Acad Sci U S A. 2012 Apr 3;109(14):5405-10. doi: 10.1073/pnas.1118801109. Epub 2012 Mar 19. PMID: 22431638; PMCID: PMC3325661.

[36] Clark SJ, Jefferies WA, Barclay AN, Gagnon J, Williams AF. Peptide and nucleotide sequences of rat CD4 (W3/25) antigen: evidence for derivation from a structure with four immunoglobulin-related domains. Proc Natl Acad Sci U S A.

1987 Mar;84(6):1649-53. doi: 10.1073/pnas.84.6.1649. PMID: 3104900; PMCID: PMC304494.

[37] Wu H, Kwong PD, Hendrickson WA. Dimeric association and segmental variability in the structure of human CD4. Nature. 1997 May 29;387(6632):527-30. doi: 10.1038/387527a0. PMID: 9168119.

[38] Chida AS, Goyos A, Robert J. Phylogenetic and developmental study of CD4, CD8 α and β T cell co-receptor homologs in two amphibian species, Xenopus tropicalis and *Xenopus laevis*. Dev Comp Immunol. 2011 Mar;35(3):366-77. doi: 10.1016/j.dci.2010.11.005. Epub 2010 Nov 21. PMID: 21075137; PMCID: PMC3073561.

[39] Claeys E, Vermeire K. The CD4 Receptor: An Indispensable Protein in T Cell Activation and A Promising Target for Immunosuppression. Arch Microbiol Immunology. 2019 3(3):133-150. doi:10.26502/ami.93650036.

[40] Ashfaq H, Soliman H, Saleh M, El-Matbouli M. CD4: a vital player in the teleost fish immune system. Vet Res. 2019 Jan 7;50(1):1. doi: 10.1186/s13567-018-0620-0. PMID: 30616664; PMCID: PMC6323851.

[41] Mørch AM, Bálint Š, Santos AM, Davis SJ, Dustin ML. Coreceptors and TCR Signaling - the Strong and the Weak of It. Front Cell Dev Biol. 2020 Oct 16;8:597627. doi: 10.3389/fcell.2020.597627. PMID: 33178706; PMCID: PMC7596257.

[42] Bowers K, Pitcher C, Marsh M. CD4: a co-receptor in the immune response and HIV infection. Int J Biochem Cell Biol. 1997 Jun;29(6):871-5. doi: 10.1016/s1357-2725(96)00154-9. PMID: 9304802.

[43] Lifson JD, Engleman EG. Role of CD4 in normal immunity and HIV infection. Immunol Rev. 1989 Jun;109:93-117. doi: 10.1111/j.1600-065x.1989.tb00021.x. PMID: 2475427.

[44] Glatzová D, Cebecauer M. Dual Role of CD4 in Peripheral T Lymphocytes. Front Immunol. 2019 Apr 2;10:618. doi: 10.3389/fimmu.2019.00618. PMID: 31001252; PMCID: PMC6454155.

[45] van der Donk LEH, Ates LS, van der Spek J, Tukker LM, Geijtenbeek TBH, van Heijst JWJ. Separate signaling events control TCR downregulation and T cell activation in primary human T cells. Immun Inflamm Dis. 2021 Mar;9(1):223-238. doi: 10.1002/iid3.383. Epub 2020 Dec 22. PMID: 33350598.

[46] Venet F, Chung CS, Huang X, Lomas-Neira J, Chen Y, Ayala A. Lymphocytes in the development of lung inflammation: a role for regulatory CD4+ T cells in indirect pulmonary lung injury. J Immunol. 2009 Sep 1;183(5):3472-80. doi: 10.4049/jimmunol.0804119. Epub 2009 Jul 29. PMID: 19641139; PMCID: PMC2788796.

[47] Kopec,AL, Michaud ZE, Maniero GD. The Role Of IL-16 As a lymphocyte attractant appears to be conserved through phylogeny: preliminary evidence that recombinant human IL-16 preferentially attracts regulatory lymphocytes in the amphibian, *Xenopus laevis*. Arch Autoimmune Dis. 2020 1(2):44-48. in press.

[48] Weiss A, Littman DR. Signal transduction by lymphocyte antigen receptors. Cell. 1994 Jan 28;76(2):263-74. doi: 10.1016/0092-8674(94)90334-4. PMID: 8293463.

[49] Huang Y, Wang W, Xu Z, Pan J, Zhao Z, Ren Q. Eriocheir sinensis microRNA-7 targets crab Myd88 to enhance white spot syndrome virus replication. Fish Shellfish Immunol. 2018 Aug;79:274-283. doi: 10.1016/j.fsi.2018.05.028. Epub 2018 May 26. PMID: 29775740.

[50] Dijkstra JM, Somamoto T, Moore L, Hordvik I, Ototake M, Fischer U. Identification and characterization of a second CD4-like gene in teleost fish. Mol Immunol. 2006 Feb;43(5):410-9. doi: 10.1016/j.molimm.2005.03.005. Epub 2005 Apr 2. PMID: 16337483.

[51] Morales H, Robert J. In vivo and in vitro techniques for comparative study of antiviral T-cell responses in the amphibian Xenopus. Biol Proced Online. 2008 Jan 17;10:1-8. doi: 10.1251/bpo137. PMID: 18385804; PMCID: PMC2275042.

[52] Buonocore F, Randelli E, Casani D, Guerra L, Picchietti S, Costantini S, Facchiano AM, Zou J, Secombes CJ, Scapigliati G. A CD4 homologue in sea bass (*Dicentrarchus labrax*): molecular characterization and structural analysis. Mol Immunol. 2008 Jun;45(11):3168-77. doi: 10.1016/j.molimm.2008.02.024. Epub 2008 Apr 9. PMID: 18403019.

[53] Moore LJ, Dijkstra JM, Koppang EO, Hordvik I. CD4 homologues in Atlantic salmon. Fish Shellfish Immunol. 2009 Jan;26(1):10-8. doi: 10.1016/j.fsi.2008.09.019. Epub 2008 Oct 15. PMID: 18983924.

[54] Picchietti S, Guerra L, Buonocore F, Randelli E, Fausto AM, Abelli L. Lymphocyte differentiation in sea bass thymus: CD4 and CD8-alpha gene expression studies. Fish Shellfish Immunol. 2009 Jul;27(1):50-6. doi: 10.1016/j.fsi.2009.04.003. Epub 2009 May 5. PMID: 19422917.

[55] Toda H, Saito Y, Koike T, Takizawa F, Araki K, Yabu T, Somamoto T, Suetake H, Suzuki Y, Ototake M, Moritomo T, Nakanishi T. Conservation of characteristics and functions of CD4 positive lymphocytes in a teleost fish. Dev Comp Immunol. 2011 Jun;35(6):650-60. doi: 10.1016/j.dci.2011.01.013. Epub 2011 Jan 25. PMID: 21272597.

[56] Wen Y, Fang W, Xiang LX, Pan RL, Shao JZ. Identification of Treg-like cells in Tetraodon: insight into the origin of regulatory T subsets during early vertebrate evolution. Cell Mol Life Sci. 2011 Aug;68(15):2615-26. doi: 10.1007/s00018-010-0574-5. Epub 2010 Nov 10. PMID: 21063894.

[57] MaiseyK,MonteroR,Corripio-MiyarY, Toro-AscuyD,ValenzuelaB,Reyes-CerpaS, Sandino AM, Zou J, Wang T, Secombes CJ, Imarai M. Isolation and Characterization of Salmonid CD4+ T Cells. J Immunol. 2016 May 15;196(10):4150-63. doi: 10.4049/jimmunol.1500439. Epub 2016 Apr 6. PMID: 27053758.

[58] Mao K, Chen W, Mu Y, Ao J, Chen X. Molecular characterization and expression analysis during embryo development of CD4-1 homologue in large yellow croaker *Larimichthys crocea*. Fish Shellfish Immunol. 2017 May;64:146-154. doi: 10.1016/j.fsi.2017.02.044. Epub 2017 Feb 27. PMID: 28254500.

[59] Somamoto T, Kondo M, Nakanishi T, Nakao M. Helper function of $CD4^{+}$ lymphocytes in antiviral immunity in ginbuna crucian carp, *Carassius auratus* langsdorfii. Dev Comp Immunol. 2014 May;44(1):111-5. doi: 10.1016/j.dci.2013.12.008. Epub 2013 Dec 14. PMID: 24342571.

[60] Pancer Z, Mayer WE, Klein J, Cooper MD. Prototypic T cell receptor and CD4-like coreceptor are expressed by lymphocytes in the agnathan sea lamprey. Proc Natl Acad Sci U S A. 2004 Sep 7;101(36):13273-8. doi: 10.1073/pnas.0405529101. Epub 2004 Aug 24. PMID: 15328402; PMCID: PMC516559.

[61] Takizawa F, Magadan S, Parra D, Xu Z, Korytář T, Boudinot P, Sunyer JO. Novel Teleost CD4-Bearing Cell Populations Provide Insights into the Evolutionary Origins and Primordial Roles of CD4+ Lymphocytes and CD4+ Macrophages. J Immunol.

2016 Jun 1;196(11):4522-35. doi: 10.4049/jimmunol.1600222. Epub 2016 May 4. PMID: 27183628; PMCID: PMC5100900.

[62] Venkatesh B, Lee AP, Ravi V, Maurya AK, Lian MM, Swann JB, Ohta Y, Flajnik MF, Sutoh Y, Kasahara M, Hoon S, Gangu V, Roy SW, Irimia M, Korzh V, Kondrychyn I, Lim ZW, Tay BH, Tohari S, Kong KW, Ho S, Lorente-Galdos B, Quilez J, Marques-Bonet T, Raney BJ, Ingham PW, Tay A, Hillier LW, Minx P, Boehm T, Wilson RK, Brenner S, Warren WC. Elephant shark genome provides unique insights into gnathostome evolution. Nature. 2014 Jan 9;505(7482):174-9. doi: 10.1038/nature12826. Erratum in: Nature. 2014 Sep 25;513(7519):574. Erratum in: Nature. 2020 Dec;588(7837):E15. PMID: 24402279; PMCID: PMC3964593.

[63] Redmond AK, Macqueen DJ, Dooley H. Phylotranscriptomics suggests the jawed vertebrate ancestor could generate diverse helper and regulatory T cell subsets. BMC Evol Biol. 2018 Nov 15;18(1):169. doi: 10.1186/s12862-018-1290-2. PMID: 30442091; PMCID: PMC6238376.

[64] Laing KJ, Zou JJ, Purcell MK, Phillips R, Secombes CJ, Hansen JD. Evolution of the CD4 family: teleost fish possess two divergent forms of CD4 in addition to lymphocyte activation gene-3. J Immunol. 2006 Sep 15;177(6):3939-51. doi: 10.4049/jimmunol.177.6.3939. PMID: 16951357.

[65] Bazan JF. Structural design and molecular evolution of a cytokine receptor superfamily. Proc Natl Acad Sci U S A. 1990 Sep;87(18):6934-8. doi: 10.1073/pnas.87.18.6934. PMID: 2169613; PMCID: PMC54656.

[66] Taga T, Kishimoto T. Cytokine receptors and signal transduction. FASEB J. 1992 Dec;6(15):3387-96. doi: 10.1096/fasebj.6.15.1334470. PMID: 1334470.

[67] Moldovan MC, Yachou A, Lévesque K, Wu H, Hendrickson WA, Cohen EA, Sékaly RP. CD4 dimers constitute the functional component required for T cell activation. J Immunol. 2002 Dec 1;169(11):6261-8. doi: 10.4049/jimmunol.169.11.6261. PMID: 12444132.

[68] Barclay AN, Brady RL, Davis SJ, Lange G. CD4 and the immunoglobulin superfamily. Philos Trans R Soc Lond B Biol Sci. 1993 Oct 29;342(1299):7-12. doi: 10.1098/rstb.1993.0129. PMID: 7904350.

Section 2

Autoimmune Diseases and Low Immune System

Chapter 3

Interleukin 6 in Patients with Rheumatoid Arthritis

Yogita Sharma, Neeraj Kumar and Devyani Thakur

Abstract

Rheumatoid Arthritis is a widespread disease causing varying degrees of disability. It is characterised by flares and remissions and since ancient times, every culture has tried to get the better of it. Even now, research is aimed at finding novel serum biomarkers as surrogates for disease activity and newer targets to sharpen therapy. One such target is IL-6.It mediates neutrophil migration, osteoclast maturation and pannus formation through vascular endothelial growth factor (VEGF) stimulation causing synovitis and joint destruction.IL-6 leads to various systemic manifestations like hepcidin production causing anemia hypothalamo-pituitary–adrenal (HPA) axis activation causing fatigue and mood changes and osteoclast activation causes osteoporosis while increase in acute phase reactants (ESR and CRP). The literature we reviewed and our research, enrolling 40 patients of RA as well describes the role of IL-6 in pathogenesis and various manifestations of RA including articular, extra-articular and other comorbid states. It supports that Serum IL-6 levels correlate with disease activity (DAS-28ESR and BRAF-MDQ) and that IL-6 remains a viable target for drug therapy.

Keywords: rheumatoid arthritis, pathophysiology IL-6, HPA axis, fatigue, DAS-28, BRAF-MDQ

1. Introduction

Rheumatoid arthritis (RA), the most common rheumatological disorder seen in clinical practice, has an estimated prevalence in the Indian community of 0.75%. It afflicts near about 1% of the world's population. Like many other connective tissue disorders, RA affects women more than men (female:male = 2:1 to 4:1) [1]. It is characterized by persistent synovial inflammation, bony erosions and progressive articular destruction, leading to varying degrees of physical disability [2]. The disease is known to produce periods of flares and remissions, therefore, it needs regular monitoring and continuous research to improve the quality of life of sufferers [3].

2. Pathogenesis of rheumatoid arthritis

RA primarily affects the musculoskeletal system which includes the synovial tissue, underlying bone and cartilage. However, being a systemic disease, it also produces a variety of extra-articular manifestations, such as subcutaneous nodules, lung involvement, peripheral neuropathy, vasculitis, pericarditis, hematological abnormalities and fatigue [4].

The macrophages are the key cells that are responsible for the tissue damage in RA. These cells are the source of pro-inflammatory cytokines involved in the pathogenesis.IL-6 is one of the main cytokine which is the cause of inflammation and immune dysregulation [4]. However, the exact pathogenic mechanism remains a complex interplay of genetic, environmental, and immunological factors that produce dysregulation of the immune system and a breakdown in self-tolerance a involvement of IL-6 and the HPA axis in the pathogenesis of fatigue has been shown in RA [5].

3. Interleukin 6 in RA

In RA, numerous cytokines, as we have already seen, are present both in the blood and in synovial joints. Hence, the cytokine network is complex and drives most of the clinical features consequently [6].

Elevation in pro-inflammatory cytokine levels leads to higher levels of fatigue in RA [7]. A significant role is being played by interleukin 6 (IL-6) in the pathogenesis of RA and promotion of fatigue [8].

4. Biology of IL-6

IL-6 is a pleiotropic cytokine. It is known to have substantial effects on non-immunological tissues [9]. It stimulates the production of acute-phase proteins, induces anemia and impairs the HPA axis [10].

Besides the immune system, this cytokine being proinflammatory causes various effects on multiple extraaticular tissues in the body which includes cardiovascular system, glucose metabolism through alteration of the insulin sensitivity, neurohormonal axis causing various psychological behavioural and haematological abnormalities [11]. Role of IL-6 is being considered in maintaining balance between immune and non immune systems of the body both in the healthy and disease states [9].

5. Molecular structure of IL-6

From a structural standpoint, IL-6 is a tetrahelix protein containing 184 amino acids [12]. It acts on various cells including leucocytes, megakaryotes and hepatocytes to name a few [11]. The receptor for IL-6 (IL-6R) are formed of an α chain, CD126, and two chains of glycoprotein 130 (gp130) [13, 14]. The signal transduction can occur through classical and trans signalling mechanisms.

In classical signalling, when IL-6 binds its membrane bound receptors and forms an IL-6/Mil-6Rα pair that leads to downstream signalling [12, 15, 18].

In trans signalling, IL-6 binds to its soluble receptor sIL-6-6Rα which further forms a complex with gp130.This IL-6/sIL-6Rα/gp130 then dimerises and leads to signal transduction [12, 15–19].

As neuronal cells prominently express gp130 and can therefore be activated via IL-6 trans-signalling, IL-6 is purported to have a direct effect on the CNS-related RA symptoms and co-morbidities, particularly, pain, fatigue, and mood [20–23].

6. IL-6 and fatigue in RA

It is well established that the cause of RA-associated fatigue is multidimensional, involving inflammation, pain, anemia, poor sleep, and psychosocial factors. There is also substantial evidence implicating the involvement of IL-6 and the HPA axis in the pathogenesis of fatigue [5].

The positive effects of IL-6 inhibition on symptoms of fatigue by Tocilizumab, Sarilumab, and Sirukumab in patients with moderate to severe RA, as assessed by FACIT-F, have been demonstrated in several clinical studies [8]. Alleviation of fatigue appears to be one of the first beneficial effects that patients with RA may experience when using biologic therapies that block IL-6 signaling [9].

This makes the precise measurement of the subjective feeling of fatigue as important and necessary as the disease activity, to evaluate the potential treatment effects [24].

Classically, the Bristol RA Fatigue Multi-Dimensional Questionnaire (BRAF-MDQ) [25] has been used for measuring fatigue in patients of RA. It was developed from the patient's perspective and evaluated in a British RA population. It was published in 2010 [25].

Nicklin et al. showed that the BRAF-MDQ global score correlated strongly with the MAF, POMS, and FACIT-F while the correlation with the SF-36 vitality subscale was weak [26].

7. Disease activity in RA

In rheumatoid arthritis, the presentation and course of the disease over time, are highly variable. The symptoms and signs of RA vary from joint complaints like pain, stiffness, swelling, and functional impairment, to more constitutional complaints like fatigue and loss of general health [4].

In the past decades a large number of variables have been tried to assess the status and course of disease activity in RA.

In daily clinical practice as well as in clinical trials on a group as well as individual level, the Disease Activity Score (DAS) and the DAS28 have been developed to measure disease activity in RA. These scores are a measure of RA disease activity that have been developed by compiling the information about swollen joints, tender joints, acute phase response, and general health. The variables required for calculation of DAS28 score include a 28-Tender joint Count (28-TJC), a 28-Swollen Joint Count (28-SJC), erythrocyte sedimentation rate (ESR), and a patient global assessment (PGA) of disease activity on a visual analog scale (VAS). C-reactive protein (CRP) may be used as an alternative to ESR in the calculation of the DAS or DAS28 [27].

Previous studies have shown that IL-6 levels were raised in the synovial membrane and synovial fluid of patients with RA [4]. However, the exact correlation of disease activity with serum IL-6 levels is still debatable in patients of RA. We did a study to measure the serum levels of IL-6 and disease activity in patients with RA and aimed to correlate the two statistically.

8. A research

Our study was conducted in the Department of Medicine between November 2016 to March 2018 in a tertiary care hospital of New Delhi. We studied 40 patients of RA (**Table 1**) who were diagnosed according to the ACR/EULAR 2010 Criteria [28].

Demographic data and disease history regarding onset, duration, course and progression, received were obtained from the patients (**Table 1**).

S.no	Character	RA patients(n = 40)
1.	Age (yrs)	38.45 ± 7.51
2.	Gender (female/male)	31/9
3.	Duration of disease	2.31 ± 1.71
4.	ESR(mm in 1st hour)	33.45 ± 20.16
5.	CRP(positive/negative)	23/17
6.	IL-6(pg/ml)	37.92 ± 75.29
7.	Rheumatoid factor(positive/negative)	29/11
8.	Anti CCP(IU/ml)	117.18 ± 107.96

Table 1.
Clinical characteristics of study population.

A general physical and thorough clinical examination of the musculoskeletal system was carried out.

DAS 28-ESR [29] was calculated for each patient as follows:

- **DAS 28 score** = 0.56 x $\sqrt{}$tender joint count +0.28 x $\sqrt{}$swollen joint count +0.70 x ln [ESR] +1.14 x (patient's global assessment on a scale of 1–100, measured using Visual analog scale).

The **cut-off values of DAS 28 for disease activity** are:

> 5.1	High disease activity,
>3.2 –≤ 5.1	Moderate disease activity
≤ 3.2-2.6	Low disease activity,
< 2.6	Remission.

Fatigue was measured using BRAF-MDQ score [25]

The data was entered in MS EXCEL spreadsheet and analysis was done using Statistical Package for Social Sciences (SPSS) version 21.0 (IBM, Chicago/USA). The normality of data was tested by the Kolmogorov–Smirnov test. Quantitative variables were compared using the Independent T-test/Mann–Whitney Test (when the data sets were not normally distributed) between the two groups and ANOVA/ Kruskal Wallis test between more than two groups. Qualitative variables were correlated using the Chi-Square test. Pearson correlation coefficient/Spearman rank correlation coefficient was used to assess the association of various parameters with each other. A p-value of <0.05 was considered statistically significant.

DAS 28 score ranged from 0.51 to 6.1 with a mean of 3.21 and a standard deviation of 1.26. The distribution of the patients by DAS28 is shown below in **Table 2.**

Total fatigue score ranged from 25 to 65 with a mean of 44.1 and ranged from minimum score of 25 to maximum score of 65.

IL-6 levels correlated with DAS28 with statistical significance, a p-value of 0.0011 and correlation coefficient of 0.497.

Chi-Square test was used to assess the correlation of DAS28 with sex and RF in the study population. But the p values of 0.240 and 0.384 respectively showed that there was no difference in disease activity between male and female patients.

According to DAS28 scores as above, patients were divided into subgroups of remission, low disease activity, moderate disease activity, and high disease activity. We studied the effect of various parameters on DAS28.

Higher concentrations of serum IL-6 were associated with higher disease activity (p = 0.0011, correlation coefficient = 0.497) as shown in **Figure 1**, however age

DAS-28		
	Frequency	**Percentage**
1) Remission(DAS28 < 2.6)	16	40.00%
2) Low disease activity(DAS28: 2.6 - ≤ 3.2)	5	12.50%
3) Moderate disease activity(Activity (DAS28: >3.2 - ≤ 5.1)	16	40.00%
4) High disease activity((DAS28 > 5.1)	3	7.50%
Total	40	100.00%

Table 2.
Distribution of disease activity by DAS28.

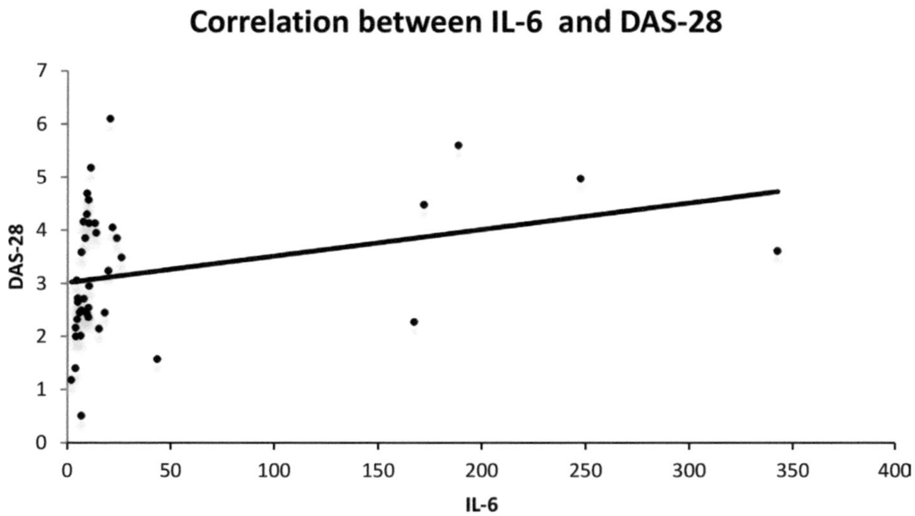

Figure 1.
Scattered plot showing correlation between IL-6 and DAS28.

(p-value = 0.1262), gender(p = 0.240), Anti CCP (p = 0.4296) and RF (p = 0.384) did not correlate with disease activity as measured by DAS28.

Levels of serum IL-6 were found to be very strongly correlating with BRAF-MDQ score, with a p-value of <0.0001 and a correlation coefficient of 0.821 as shown in **Figure 2**.

In our study, the levels of serum interleukin-6(IL-6) in the patients were high with a mean of 37.92 ± 75.29 pg/ml and ranged from 1.95 to 342.5 pg/ml. This finding was consistent with the results of other studies done previously. In the study done by Helal et al. [30] serum IL-6 concentration was significantly elevated in patients with RA ranging between 5 and 130 pg/ml, with a mean of 35.0 ± 21.2.

In a study done by Chung et al. [31] on the correlation between increased serum concentration of IL-6 family cytokines and disease activity in rheumatoid arthritis, the serum concentrations of IL-6 were 41.76 ± 20.28 pg/ml (range:18.0 to 109.1 pg/ml).

IL-6 is one of the cytokines which play a significant role in the pathogenesis of RA and the promotion of fatigue [6, 10, 32–34].Analytical statistics were also done to assess the correlation between BRAF-MDQ score and serum IL-6. Levels of serum IL-6 were found to be very strongly correlated with the BRAF-MDQ score with a p-value of 0.0001 and a correlation coefficient of 0.821. Our results were comparable to those of. Helal et al. [30] They too, found a strong correlation between BRAF-MDQ score and serum IL-6 concentration with r = 0.947, p < 0.001.

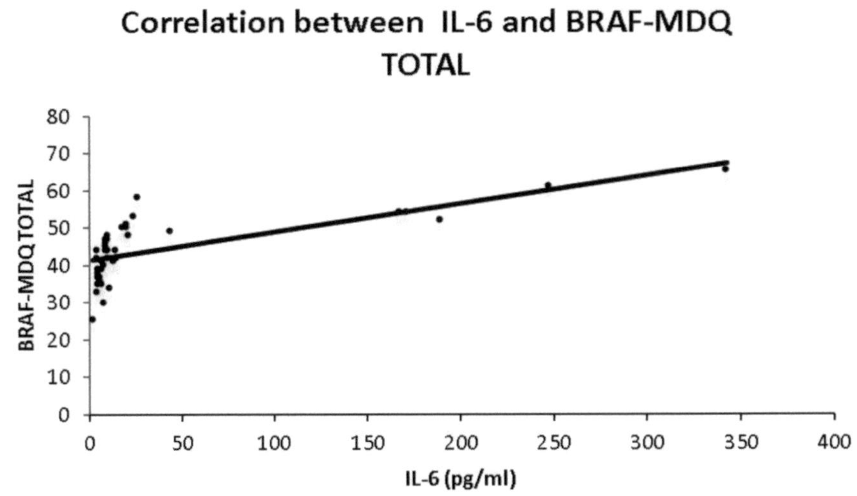

Figure 2.
Correlation between IL-6 and BRAF-MDQ score.

9. IL-6 in various diseases

Environmental stress factors such as infections and tissue injuries trigger immediate and transient rise in the levels of IL-6 which activates host defense mechanisms. As this stress is removed from the host the signal transduction and inflammatory cascade are terminated [35].

Dysregulated IL-6 production leads to the development of various immune and non-immune mediated diseases [35]. This was first demonstrated in a case of cardiac myxoma and remains true till date as seen in the COVID-19 pandemic.

A study done by Hirano et al. [36] in 1988 showed that dysregulation of IL-6 production occurs in the synovial cells of RA. Various gene knockout studies and IL-6 blockade by administration of anti-IL-6 or anti-IL-6R Ab have shown to be promising in the prevention and alleviation of disease symptoms [6, 8]. Mitigation of disease symptoms by this strategy has been shown by Alonzi et al. [37] Ohshima et al. [38] Fujimoto et al. [39] in patients with rheumatoid arthritis.

10. IL-6 and its systemic effects in SRA

In inflammatory arthritis, Osteoclasts play a major role in causing bony erosions [40]. Osteoclasts are recruited by IL-6 that acts on hematopoietic stem cells from the granulocyte-macrophage lineage (**Figure 3**) [41–43].

IL-6 has also been recognized to play a major role in extracellular matrix turnover and levels of IL-6 and CRP correlate with proMMP-3 in patients with early RA [44] which shows a link between IL-6 and proteinase activity. It stimulates hepatocytes to increase the production of acute-phase reactants. The correlation of IL-6 with CRP is seen in RA patients [10].

The anemia of chronic inflammations is seen in RA patients. The iron transport and release of iron from macrophages are inhibited by protein hepcidin which is produced by hepatocytes [10, 45].

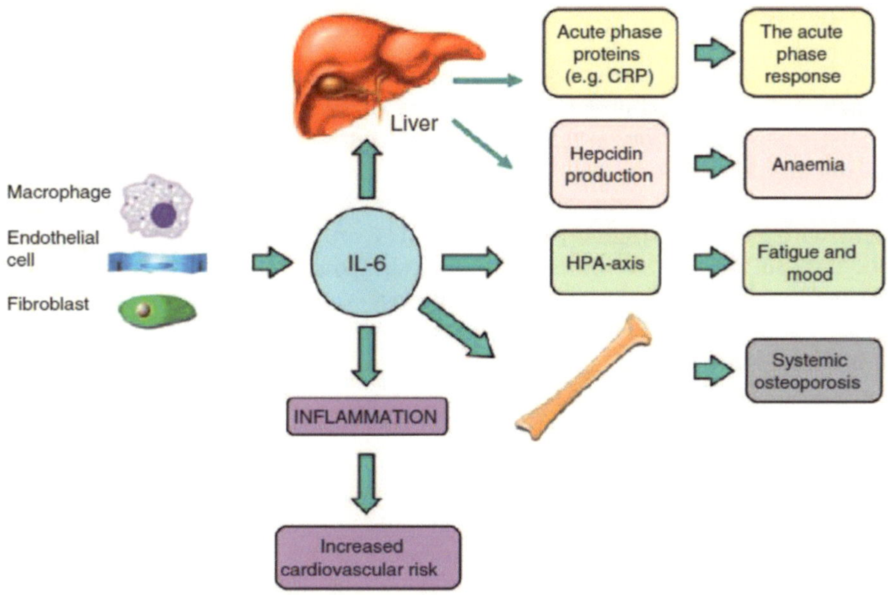

Figure 3.
Systemic effects of IL-6.

The hepcidin regulates iron metabolism by preventing iron transport and the release of iron from macrophage [45].

One of the common systemic manifestations of RA is osteoporosis. IL-6 over-expression results in osteopenia due to osteoclast and osteoblast dysregulation. This was shown in in-vivo studies with IL-6 transgenic mice resulting in increased osteoclastogenesis that leads to accelerated bone resorption, reduced bone formation, and defective ossification [46].

Cardiovascular mortality is predominant in patients with RA. In RA, endothelial dysfunction and dyslipidemia lead to an increased risk of atherogenesis because of systemic inflammation [47–49]. The widespread systemic inflamation is proportionate to elevated CRP levels which leads to increased risk of cardiovascular disease [50].

11. IL-6 blockade as a therapeutic target in RA

As IL-6 has been shown to have an array of biological roles and pathological efsfects in immune diseases, IL-6 targetting would constitute a novel therapeutic option in RA as well. This has been shown in the OPTION study [8] where Tocilizumab has been shown to reduce diseases activity and led to improvement in all ACR core set variables when compared with patients who received placebo(less than 1% on placebo—achieved DAS28 remission). The physical disability was substantially reduced by Tocilizumab more as compared to placebo, suggesting considerable functional benefits for the patients. Also Tocilizumab lead to more improvements in health-related quality of life than with placebo. Sustained improvements in the acute phase response markers including ESR, CRP and, and hemoglobin, were seen, especially with tocilizumab 8 mg/kg. In TAMARA study s, Tocilizumab was highly effective in a setting close to real-life medical care with a rapid and sustained improvement in signs and symptoms of RA [51].

12. Conclusion

It is well established that synovial cytokines, particularly IL-6 are responsible for much of the destruction in RA. The review also suggests that IL-6 is involved in the pathogenesis of various extra-articular manifestations of rheumatoid arthritis including increased risk of cardiovascular diseases, deranged glucose and lipid metabolism and various neurohormonal and psychological behavioural changes in patients with RA. Even, high levels of *serum* IL-6 are associated with a high disease activity, as indicated by various studies, including ours ($p = 0.0011$, correlation coefficient = 0.497). Also, we found that the levels of serum IL-6 very strongly correlated with fatigue, as measured by the BRAF-MDQ score.

It is thus, evident that blocking the IL-6 pathway as a therapeutic target in patients with rheumatoid arthritis, may help in better control of the disease symptoms and prevent flares. The extra-articular manifestations can also be controlled by antagonising IL-6 activity.

So, in conclusion, serum IL-6 is one of the main cytokine that has been involved in the pathophysiology of RA through its complex signalling pathways and as its levels correlate with disease activity, it has emerged as a better test for measuring disease remission and flares. It is simple, convenient and gives a lucid, objective value to a largely subjective and complicated issue in the course of RA-disease activity. And therefore, IL-6 can also prove to be a novel therapeutic target in control of articular as well as extra-articular manifestations of Rheumatoid arthritis.

Author details

Yogita Sharma[1*], Neeraj Kumar[2] and Devyani Thakur[1]

1 Dr. RML Hospital, New Delhi, India

2 Civil Hospital Bhoranj, Hamirpur, India

*Address all correspondence to: dryogita12@gmail.com

References

[1] Alam SM, Kidwai AA, Jafri SR. Jour Pak med assoc. Epidemiology of rheumatoid arthritis in a tertiary care unit. Karachi, Pak. J Pak Med Assoc. 2011; 61(2):123-126.

[2] Sousa EV, Danielle M, Gerlag DM, Paul P. Synovial Tissue Response to Treatment in Rheumatoid Arthritis. Open Rheumatol J. 2011; 5:115-122.

[3] Saleem B, Brown AK, Quinn M, Karim Z, Hensor EM, Conaghan P, et al. Can flare be predicted in DMARD treated RA patients in remission, and is it important? A cohort study. Ann Rheum Dis.2012; 71:1316-1321.

[4] .AnkoorShah,William St. Clair E.Rheumatoid Arthritis. In: Kasper D. Harrison's Principles of Internal Medicine. Fauci AS, Hauser SL, Longo DL, Jameson JL, Loscalzo J, editors. Volume 2,19th edition. McGraw-Hill Medical Publishing Division; 2016 May 22;2136-2149

[5] ChrousosGP.The hypothalamic–pituitary–adrenal axis and immune-mediated inflammation. N Engl J Med 1995;332:1351-1362.

[6] Garnero P, Thompson E, Woodworth T, Smolen JS. Rapid and sustained improvement in bone and cartilage turnover markers with the anti-interleukin-6 receptor inhibitor tocilizumab plus methotrexate in rheumatoid arthritis patients with an inadequate response to methotrexate: results from a substudy of the multicesnter double-blind, placebo-controlled trial of tocilizumab in inadequate responders to methotrexate alone. Arthritis Rheum 2010;62:33-43.

[7] Dayer E, Dayer J.M, Roux-Lombard Primer P: the practical use of biological markers of rheumatic and systemic inflammatory diseases Nat ClinPractRheumatol 2007; 3:512-520.

[8] Smolen JS, Beaulieu A, Rubbert-Roth A, Ramos-Remus C, Rovensky J, Alecock E, Woodworth T, Alten R, OPTION Investigators. Effect of interleukin-6 receptor inhibition with tocilizumab in patients with rheumatoid arthritis (OPTION study): a double-blind, placebo-controlled, randomised trial. The Lancet. 2008 Mar 22;371(9617):987-997.

[9] Rohleder N, Aringer M, BoentertM. Role of interleukin-6 in stress, sleep, and fatigue. Ann N Y AcadSci 2012;1261:88-96

[10] S. Srirangan, E.H. Choy. The role of Interleukin 6 in the pathophysiology of rheumatoid arthritis. TherAdvMusculoskel Dis 2010; 2:247-256.

[11] Hunter CA, Jones SA.IL-6 as a keystone cytokine in health and disease Nat Immunol 2015;16:448-457.

[12] Rose-John S.IL-6 trans-signaling via the soluble IL-6 receptor: importance for the pro-inflammatory activities of IL-6. *Int J Biol Sci*2012;8:1237-1247.

[13] Yin T, Taga T, Tsang ML et al. Involvement of IL-6 signal transducer gp130 in IL-11-mediated signal transduction. J Immunol 1993;151:2555-2561.

[14] Yamasaki K, Taga T, Hirata Y et al. Cloning and expression of the human interleukin-6 (BSF-2/IFN beta 2) receptor. Science 1988;241:825-828

[15] Narazaki M, Witthuhn BA, Yoshida K et al. Activation of JAK2 kinase mediated by the interleukin 6 signal transducer gp130. Proc Natl AcadSci USA 1994;91:2285-2289.

[16] Scheller J, Chalaris A, Schmidt-Arras D, Rose-John S.The pro- and anti-inflammatory properties

of the cytokine interleukin-6. BiochimBiophysActa 2011;1813:878-888.

[17] Zhong Z, Wen Z, Darnell JE Jr. Stat3: a STAT family member activated by tyrosine phosphorylation in response to epidermal growth factor and interleukin. Science1994;264:95-8.

[18] Scheller J, Garbers C, Rose-John S.Interleukin-6: from basic biology to selective blockade of pro-inflammatory activities. SeminImmunol 2014;26:2-12.

[19] O'Shea JJ, Gadina M, Schreiber RD.Cytokinesignaling in 2002: new surprises in the Jak/Stat pathway. Cell 2002;109(Suppl):S121–S131.

[20] Jostock T, Müllberg JG, Özbek S et al. Soluble gp130 is the natural inhibitor of soluble interleukin-6 receptor transsignaling responses. Eur J Biochem 2001;268:160-167.

[21] Mihara M, Hashizume M, Yoshida H, Suzuki M, Shiina M.IL-6/IL-6 receptor system and its role in physiological and pathological conditions. ClinSci 2012;122:143-159.

[22] Eijsbouts AM, van den Hoogen FH, LaanRFet al. Hypothalamic-pituitary-adrenal axis activity in patients with rheumatoid arthritis.*ClinExpRheumatol* 2005;23:658-64.

[23] Schaible H-G.Nociceptive neurons detect cytokines in arthritis. Arthritis Res Ther 2014;16:470.

[24] Hewlett S, Dures E, Almeida C. Measures of Fatigue. Arthritis Care Res 2011; 63(S11):S263-S286

[25] Nicklin J, Cramp F, Kirwan J, Greenwood R, Urban M, Hewlett S.Measuring fatigue in RA: a cross-sectional study to evaluate the Bristol Rheumatoid Arthritis Fatigue Multi-Dimensional Questionnaire,Visual Analog Scales, and Numerical Rating Scales. Arthritis Care Res 2010;62:1559-68.

[26] Nicklin J, Cramp F, Kirwan J, Urban M, Hewlett S. Collaboration with patients in the design of patient-reported outcome measures: capturing the experience of fatigue in rheumatoid arthritis. Arthritis Care Res 2010;62:1552-8.

[27] Fransen J, Welsing PMJ, De Keijzer RMH et al. Development and validation of the DAS28 using CRP. Ann Rheum Dis 2003;62 (Suppl. 1):10.

[28] Aletaha D, Neogi T, Silman AJ, et al. 2010 Rheumatoid arthritis classification criteria: an American College of Rheumatology/European League Against Rheumatism collaborative initiative. Arthritis Rheum. 2010; 62:2569-2581.

[29] Prevoo ML, van't Hof MA, Kuper HH, et al. Modified disease activity scores that include twenty-eight-joint counts. Development and validation in a prospective longitudinal study of patients with rheumatoid arthritis. Arthritis Rheum 1995; 38:44-48.

[30] Abdel Moneim H. Helala, Enas M. Shahinea, Marwa M. Hassana, Doaa I. Hashadb, Riham Abdel Moneima. Fatigue in rheumatoid arthritis and its relation to interleukin-6 serum level. The Egyptian Rheumatologist 2012; 34(4):153-157.

[31] Chung SJ, Kwon YJ, Park MC, Park YB, Lee SK. The correlation between increased serum concentrations of interleukin-6 family cytokines and disease activity in rheumatoid arthritis patients. Yonsei Med J 2011;52:113-120.

[32] M.Bax, J. van Heemst, T.W. Huizinga, R.E. Toes Genetics of rheumatoid arthritis: what have we learned? Immunogenetics 2011; 63:459-466

[33] Emonts, M.J. Hazes, J.J. Houwing-Duistermaat, C.E. De Jongh, L. De Vogel, H.K. Han, *et al.* Polymorphisms in genes controlling inflammation and tissue repair in rheumatoid arthritis: a case control study BMC Med Genet 2011; 12: 36

[34] Milman N, Karsh J, Booth RA. Correlation of a multi-cytokine panel with clinical disease activity in patients with rheumatoid arthritis. ClinBiochem 2010;43:1309-1314.

[35] Naka T, Narazaki M, HirataM, Matsumoto T, Minamoto S, Aono A, Nishimoto N, Kajita T, Taga T, Yoshizaki K, et al.1997. Structure and function of a new STAT-inducedSTAT inhibitor. Nature 387: 924-929.

[36] Hirano T, Matsuda T, Turner M, Miyasaka N, Buchan G,Tang B, Sato K, Shimizu M, Maini R, Feldmann M, et al.1988. Excessive production of interleukin 6/B cell stimulatoryfactor-2 in rheumatoid arthritis. Eur J Immunol18: 1797-1801.

[37] Alonzi T, Fattori E, Lazzaro D, Costa P, Probert L, Kollias G, De Benedetti F, Poli V, Ciliberto G. 1998. Interleukin 6 isrequired for the development of collagen-induced arthritis. J Exp Med 187: 461-468.

[38] Ohshima S, Saeki Y, Mima T, Sasai M, Nishioka K, Nomura S, Kopf M, Katada Y, Tanaka T, Suemura M, et al. 1998.Interleukin 6 plays a key role in the development of antigen-induced arthritis. Proc Natl AcadSci 95: 8222-8226.

[39] Fujimoto M, Serada S, Mihara M, Uchiyama Y, Yoshida H, KoikeN, Ohsugi Y, NishikawaT, Ripley B, Kimura A, et al.2008. Interleukin-6 blockadesuppressesautoimmunearthritis in mice by the inhibition of inflammatoryTh17 responses. Arthritis Rheum 58: 3710-3719.

[40] Walsh, N.C., Crotti, T.N., Goldring, S.R. and Gravallese, E.M. (2005) Rheumatic diseases: theeffects of inflammation on bone. Immunol Rev208: 228_251.

[41] Yoshitake, F., Itoh, S., Narita, H., Ishihara, K. and Ebisu, S. (2008) Interleukin-6 directly inhibits osteoclastdifferentiation by suppressing receptor activatorof NF-kappaBsignaling pathways. J BiolChem283: 11535_11540.

[42] Liu, X.H., Kirschenbaum, A., Yao, S. and Levine, A.C. (2005) Cross-talk between the interleukin-6 andprostaglandin E(2) signalling systems results inenhancement of osteoclastogenesis through effects onthe osteoprotegerin/receptor activator of nuclearfactor-{kappa}B (RANK) ligand/RANK system.Endocrinology 146: 1991_1998

[43] Otsuka, T., Thacker, J.D. and Hogge, D.E. (1991) The effects of interleukin 6 and interleukin 3 on early hematopoietic events in long-term cultures of human marrow. ExpHematol 19: 1042_1048

[44] Roux-Lombard, P., Eberhardt, K., Saxne, T., Dayer,J.M. and Wollheim, F.A. (2001) Cytokines, metalloproteinases,their inhibitors and cartilage oligomericmatrix protein: relationship to radiological progressionand inflammation in early rheumatoid arthritis. A prospective 5-year study. Rheumatology 40: 544_551.

[45] Ganz, T. (2003) Hepcidin, a key regulator of ironmetabolism and mediator of anemia of inflammation. Blood 102: 783_788.

[46] De Benedetti, F., Rucci, N., Del, F.A., Peruzzi, B.,Paro, R., Longo, M. et al. (2006) Impaired skeletaldevelopment in interleukin-6-transgenic mice: amodel for the impact of chronic inflammation onthe growing skeletal system. Arthritis Rheum54: 3551_3563.

[47] van Leuven, S.I., Franssen, R., Kastelein, J.J., Levi, M., Stroes, E.S. and Tak, P.P. (2008) Systemicinflammation as a risk factor for atherothrombosis. Rheumatology (Oxford) 47: 3_7.

[48] Dessein, P.H., Norton, G.R., Woodiwiss, A.J., Joffe, B.I. and Wolfe, F. (2007) Influence of nonclassical cardiovascular risk factors on the accuracy of predictingsubclinical atherosclerosis in rheumatoid arthritis. JRheumatol 34: 943_951.

[49] Niessner, A., Goronzy, J.J. and Weyand, C.M. (2007) Immune-mediated mechanisms in atherosclerosis:prevention and treatment of clinical manifestations. Curr Pharm Des 13: 3701_3710.

[50] Yeh, E.T. (2004) CRP as a mediator of disease.Circulation 109: 1111_1114.

[51] BurmesterGR, Feist E, Keller H. *et al.* Effectiveness and safety of the interleukin 6- receptor antagonist tocilizumab after 4 and 24 weeks in patients with active rheumatoid arthritis:the first phase IIIb real-life study (TAMARA0)Ann Rheum Dis.2011; 70:755-759.

Chapter 4

Therapeutic Potential of IL-9 in Allergic and Autoimmune Diseases

Ahmed Ummey Khalecha Bintha,
Amani Souwelimatou Amadou, Mursalin Md Huzzatul
and Muhammad Fauziyya

Abstract

Interleukin-9 (IL-9) is a pleiotropic cytokine produced by several immune and epithelial cells. Recently, many studies have eluded the physiological and pathological roles of IL-9 and its lineage-specific helper T cell subset (Th9). In this chapter, we will focus on the immunological role of Interleukin 9 (IL-9) in allergy and autoimmunity. We will introduce the basics of IL-9 and describe the cells involved in the secretion, signaling, and regulation of IL-9. After establishing the background, we will discuss the pathogenesis and regulation of IL-9 in allergic and autoimmune diseases. We will conclude the chapter by providing an updated therapeutics that target IL-9 and their potential uses in autoimmune and allergic diseases.

Keywords: IL-9, Th9, multiple sclerosis, Th17, IBD, uveitis, mast cells, asthma, atopic dermatitis, food allergy, diabetes, TGF-β, ILC2

1. Introduction

Interleukin-9 (IL-9) is a pleotropic cytokine that regulates diverse immunological functions (**Figure 1**). This cytokine was first identified in the late 1980s as a T cell growth factor [1]. Because of the molecular weight of IL-9, it was initially known as P40 [2]. Later studies revealed that the observed molecular weight was due to N-link glycosylation, and actual molecular weight for this discovered molecule is 14 kDa [3]. A similar factor was also identified from Th2 cells and mast cells where it was initially named as T-cell growth Factor III (TCGF III) and mast cell growth-enhancing activity (MEA), respectively [2, 4]. Further studies revealed that both TCGF III and MEA actually represent the P40 factor [4]. In later years, considering its pleotropic roles and the redundant nomenclature the P40 factor was renamed as IL-9 [5].

The locus encoding IL9 in mouse is about 11 kb in size, and located on chromosome 13 [6]. The Il9 locus is comprised of 5 exons and 4 introns [3]. The Il9 locus encode for a precursor peptide of 144 amino acids, first 18 amino acids of which is signal sequence peptide. The mature IL-9 peptide, a single-chain glycoprotein of 126 amino acids, and similar to other cytokines of IL-2 family folds into a four-alpha-helix bundles [7]. Human IL-9 locus is present on chromosome 5 in the region q31–35 [6]. Homology between mouse and human IL-9 is about 55%, and both of them contain a conserved 10 cysteine residue to form a disulfide bond that is critical for a mature IL-9 peptide. Interestingly, three conserved non-coding sequences,

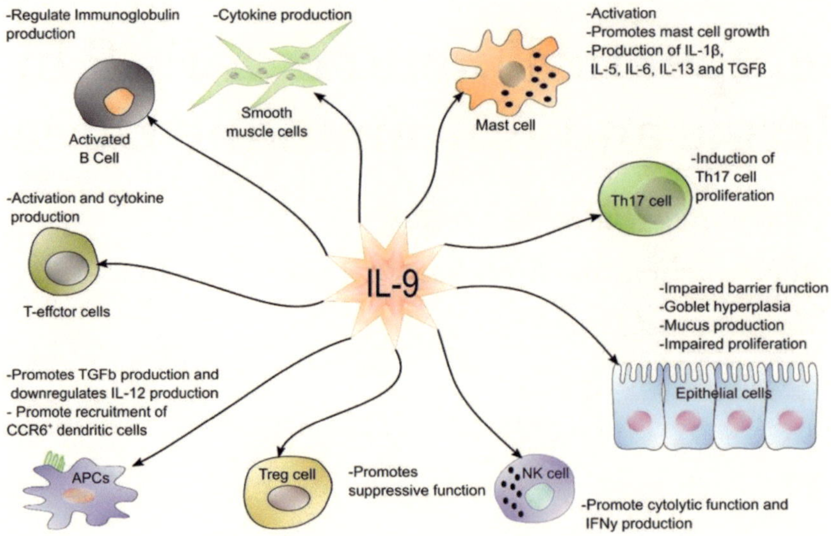

Figure 1.
Functions of IL-9. IL-9 contributes to different immunopathology and physiology through activation of multiple cell types. Illustration by MHuzzatul.

CNS0, CNS1, and CNS2 are present on both mouse and human *il9* locus sequence similarity of which is 63% [3, 7]. CNS0 is positioned in the upstream (−6 kb) of transcription start site (TSS), CNS1 is the promoter region, and CNS2 is located at the downstream of TSS (+5.4 kb) [8]. CNS1 provide binding site to numerous transcription factors that includes PU.1, STAT5, STAT6, GATA1, GATA3, IRF1, IRF4, NF-kb, BATF, AP-1, Smads 2/3/4, Gcn5, Notch [9]. Etv5 can bind to both CNS0 and CNS2, and recruit histone acetyltransferase p300 to mediate chromatin remodeling [8–11]. Regulation of IL-9 expression by this multiple numbers of transcription factors explain the necessity of a delicate cytokine milieu that requires to stimulate IL-9 producing cells. The miscellaneous origin of IL-9 and the complexity of its regulation underscore the need for a comprehensive assessment of IL-9 function. Therefore, in this chapter, we will elucidate the basis of IL-9 function in health and diseases and its therapeutic potentials in autoimmune and allergic diseases.

2. IL-9, a lineage specific Th9 cytokine

T cells were originally thought to be the main source of IL-9 [12–14]. IL-9 was defined as a Th2 cytokine. The reason for this Th2 designation by many research findings included IL-9 genome. The *il-9* gene is positioned within a Th2 cytokine clusters. Also, increased expression of IL-9 was observed in a Th2-predominate BALB/c mouse model of cutaneous leishmaniasis (BALB/c mice) but not in Th1-predominate model (using C57BL/6 mice). This finding suggested IL-9 as a Th2 signature cytokine [12]. In addition, Th2-like responses such as airway epithelial hyperplasia, proliferation of mast cells, mucin-producing cells, and eosinophils were found in the lungs of IL-9 transgenic mice [15]. More recently, the designation of IL-9 as a Th2 cytokine loses credence, due to the identification of PU.1, an ETS family transcription factor that induces IL-9 secretion. Mice with T-cell-specific deletion of PU.1 did not develop IL-9 dependent inflammation of the lungs [16]. However, the mice had similar frequencies of Th2 cells [16]. In another experiment that utilized siRNA-mediated disruption of PU.1 resulted in impaired IL-9 production in human T-cells. Recently, a distinct helper T cell subset, Th9 was identified as IL-9 lineage-specific cells. Studies observed increased PU.1 expression under Th9 polarizing conditions but not Th2 conditions [16]. The finding of another helper T cell subset suggested that Th2 is not the main source for IL-9, and PU.1 as a unique transcription factor necessary for IL-9 production emphasized the identity of Th-9. Later, *in vitro* studies identified IL-4 and TGF-β as cytokines that facilitate the differentiation of naïve T cells to Th9 cells [17, 18]. Though IL-4 is a known Th2 cytokine, TGF-β exhibit pleotropic functions and regulates the development of other helper T cells including Th17 and Treg cells [19]. Presence of IL-4 with TGF-β facilitates the differentiation of naive T cells into IL-9-secreting Th9 but not Tregs or Th17. Also, IL-4 can directly block the expression of FoxP3 in T cells thus reprogramming Treg cells into Th9 cells [17]. And, addition of TGF-β in culture medium reprograms Th2 cells to Th9 cells [18]. IL-4 and TGF-β-mediated induction of IL-9-producing cells are dependent on both activated STAT6 and GATA3, suggesting the initial identification of IL-9 as a Th2 cytokine. And Th2 including other helper T cells secrete small amounts of IL-9 [20].

3. Sources of IL-9

In addition to Th9 and Th2, other immune cells have been identified as potential sources of IL-9 (**Figure 2**). Prominent among these immune cells is Th17 cells. Th17 cells are involved in mounting immune responses against extracellular bacteria and fungi and are implicated in autoimmunity [21]. Activation of a Th17-associated transcription factor, retinoic acid receptor-related orphan receptor-γt (RORγt) with phorbol 12-myristate 13-acetate and ionomycin (PMA) leads to IL-9 secretion [22]. Tregs have also be shown to secrete IL-9 both *in vivo* and *in vitro*, however, the role is IL-9-secreting Tregs is conflicting [23, 24]. Another recently identified source of IL-9 is Vδ2 T cells in human peripheral blood. This γδ T cell subset population can be stimulated with antigens, TGF-β, and IL-15 to produce IL-9 [24]. Mast cells, natural killer T cells (NKT) have also been found to produce IL-9. Mast cells cross-linked with IgE and inflammatory mediators like histamine produce IL-9 in the presence of IL-1β and LPS [25–29]. Stimulation of NKT cells with IL-2 leads to secretion of IL-9 [30]. A large number of infiltrating IL-9 producing NKT has been found in histological section from patient with nasal NKT cell lymphomas [31]. Decreased expression of IL-9 was observed in CD1d-restricted NKT deficient mouse model of allergic inflammation suggesting NKT cell can also promote IL-9 production *in vivo* [32]. In addition, innate lymphoid cells such as ILC2s, eosinophils, neutrophils, and osteoblasts also have been found to produce IL-9 [33–35].

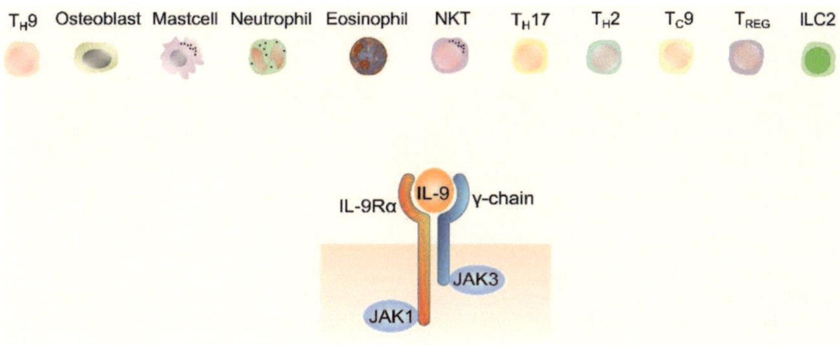

Figure 2.
Cellular sources of IL-9 and IL-9 receptor (IL-9R) heterocomplex. Illustration by MHuzzatul.

4. IL-9 receptor signaling

IL-9 exerts its biological effect on its target cells through IL-9R receptor. The IL-9R is a heterocomplex of the alpha chain (IL-9Rα) and the common gamma chain [36]. IL-9Rα is specific only to IL-9, whereas the gamma chain is present in the receptor complexes of several other cytokines such as IL-2, IL-4, IL-7, IL-13, IL-15, and IL-21 [37–39]. About 25% of the IL-9Rα exist in complex with the gamma chain outside IL-9 heterocomplex. IL-9Rα is of 522 amino acids in human, and 468 amino acids in mouse, and contains 11 exons [40]. This 64 kDa glycoprotein is a member of type I hematopoietin receptor super family due to the presence of the Box1 and Box2 motifs in the intracellular domain, and WSXWS motif in the extracellular domain [41]. Formation of a heterocomplex with the γ-chain is enhanced as IL-9 binds to IL-9Rα (**Figure 2**) [42]. The binding of IL-9 to IL-9Rα results in a conformational change in IL-9R. This conformational change recruit JAK molecules to Box1 motif which results in the phosphorylation of tyrosine residues of IL-9Rα-associated JAK1 and γ-chain associated JAK3 [41]. BOX1 motif is very critical in IL-9 mediated signaling as disruption of Box1 results in loss of

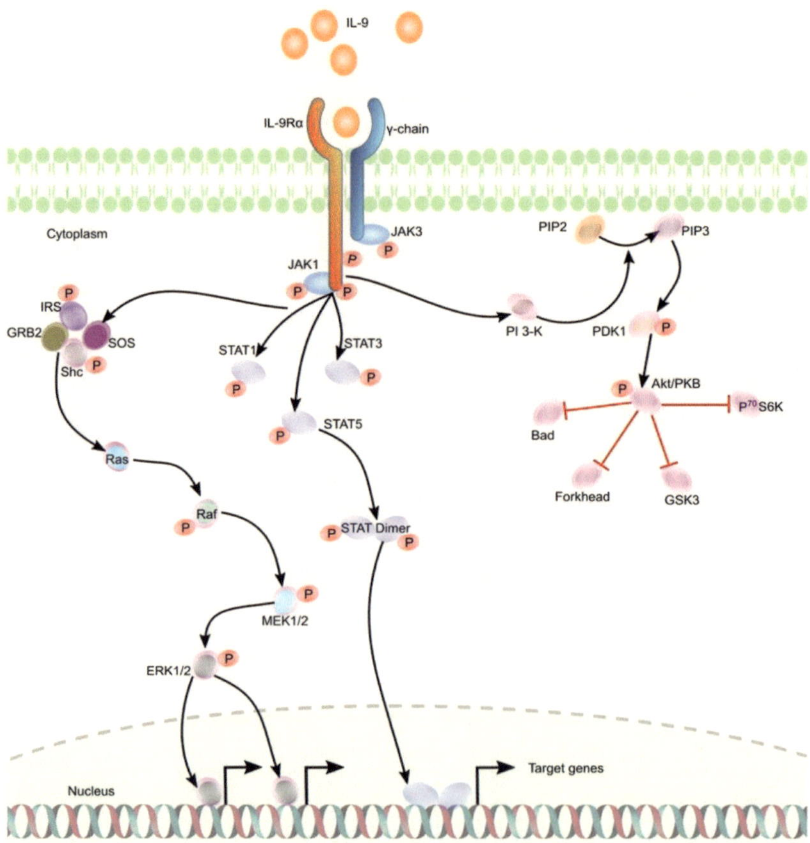

Figure 3.
Schematic representation of IL-9 signaling pathway. IL-9 cytokine binds to IL-9R complex. This leads to phosphorylation of JAKs. The phosphorylated JAKs activate STATs, PI3 kinase, and the MAP kinase pathway. IL-9R, interleukin-9 receptor; JAK, Januse kinase; STAT, signal transducer and activator of transcription; PI3K, phosphatidylinositol-3 kinase; PIP, phosphoinositide; PDK1, pyruvate dehydrogenase kinase 1; bad, GSK3, glycogen synthase kinase 3; PS6K, IRS, insulin receptor substrate; SOS, suppressors of cytokine signaling; GRB2, ERK, extracellular signal regulated kinase; Shc; Ras/Raf/MEK, mitogen-activated protein kinases; illustration by MHuzzatul.

phosphorylation of JAK1 and JAK3 [43]. Activated JAK molecules then phosphorylate a tyrosine residue (Tyr407) in the IL-9Rα, which results in the phosphorylation of intermediate molecules, STAT molecules (STAT1, STAT3, and STAT5), MAPK, and IRS-PI3 pathways (**Figure 3**) [44–46]. Activation of these pathways contribute to the upregulation of IL-9, as well as important in the growth, differentiation, and development of the IL-9 targeted cells [47, 48].

5. IL-9 and allergic diseases

Allergic diseases including respiratory, food, and skin allergies are mainly mediated by Th2 cells through the expression of various cytokines such as IL-4, IL-5, and IL-13 (reviewed in [49]). The cytokine IL-9, which was initially studied in the context of Th2-mediated immune response and later associated with T-helper 9 (Th9) cells, has been shown to play an important role in allergic inflammation [50, 51]. IL-9 and its receptor IL-9Rα regulate antibody synthesis, specifically IgE, in both murine and human B cells [52, 53]. To contribute to allergic disease pathogenesis, IL-9 also promotes activation and recruitment of inflammatory cells [54–57].

6. Asthma including airway allergies

Various studies have shown that IL-9 and its receptor contribute to airway allergic diseases and asthma. Sputum, serum, and lungs of patients with asthma were shown to have increased concentrations of the cytokine [58–60]. IL-9 levels were also increased in the airways of murine asthma models [61]. IL-9Rα is expressed on human tonsillar germinal center and memory B cells, and smooth muscles in the airways. IL-9/ IL-9Rα signaling in B cells induces STAT3 and STAT5 pathways to potentiate IgE production [52, 53, 55, 62, 63]. Overexpression of IL-9 in transgenic mice or treatment with recombinant cytokine induces expansion of B-1 cells, and accumulation of mast cells in the tissues [64, 65]. IL-9 induces the release of proteases and pro-inflammatory cytokines by the mast cells to promote survival of eosinophils and increase airway permeability [66, 67]. IL-9/IL-9Rα signaling also stimulates human airway smooth muscle to secrete eotaxin1/CCL1 and induces production of IL-13 in airway epithelial cells. Eotaxin1/CCL11 and IL-13 significantly increase eosinophil recruitment and cause lung epithelial cell hypertrophy. These effects result in asthma-like symptoms, including lung inflammation, bronchial hyper-responsiveness, and mucus accumulation. Moreover, IL-9 worsens lung injury in a murine model of chronic obstructive pulmonary disease (COPD) [63, 68, 69]. The cytokine also appears to be a critical player in allergic rhinitis. Serum IL-9 in patients strongly correlates with irritative nasal symptoms including rhinorrhea [70]. In mice, Th9 cells are significantly upregulated during allergic rhinitis and neutralization of IL-9 alleviates symptoms. Blocking IL-9 decreases the level of inflammatory cytokines (IFN-γ, IL-4, and IL-17) and eosinophils infiltration in the nasal mucosa. This causes a decrease in the frequency of sneezing and nasal rubs in experimental models of allergic rhinitis [71].

7. Food allergies

Studies in patients with food allergy and experimental oral hypersensitivity have shown that allergic reactions in the gastrointestinal tract are mediated by various players, including Th2-secreted cytokines, such as IL-4 and IL-9 [72–74]. Various studies have shown that IL-9 drives intestinal inflammation and plays a critical role in food allergies [75, 76]. In patients with food allergies, the severity of clinical symptoms strongly correlates with increased intestinal permeability [77]. *In vitro* experiments have shown that patients with peanut allergy have increased levels of IL-9. The memory T helper cell response specific to peanuts in allergic children is dominated by IL-9. Thus, cytokine levels can be used as a biomarker to determine individuals with peanut allergy [78, 79]. In mice, overexpression of intestinal IL-9 or induction of IL-9-producing mucosal mast cells (MMC9s) also increases susceptibility to food allergy [80]. Migration of mast cell progenitors and their development into MMC9s is regulated by basic leucine zipper transcription factor ATF-like (BATF) and Th2-secreted IL-4 [81]. The large amount of MCC9s-derived IL-9 and other mast cell mediators cause intestinal mastocytosis and increased intestinal permeability, which is central to the induction of experimental oral hypersensitivity [82]. The actions of the IL-9-stimulated mast cells cause allergic diarrhea and hypothermia [75]. IL-9 can additionally be secreted by the group 2 innate lymphoid cells (ILC2) and Th9 cells to amplify the intestinal allergic inflammatory response, which may lead to anaphylaxis [83–88].

8. Skin allergies

IL-9 has been identified as a potential mediator of cutaneous allergies, including atopic dermatitis (AD) and allergic contact dermatitis (ACD). Patients with atopic dermatitis have a significantly higher level of IL-9 in the serum and skin lesions [89]. The concentration of the cytokines also positively correlates with the severity of the disease and serum IgE levels [90]. These observations were made in both adult and pediatric patients [91, 92]. A study in a Korean population also linked IL-9 and IL-9R gene polymorphisms to AD [93]. IL-9 induces IL-5 and IL-13 by ILC2. ILC2 and the cytokines are associated with AD pathogenesis. IL-5 and IL-13 contribute to the defective skin barrier in AD patients by downregulating tight junctions genes [94, 95]. IL-9 also promotes the secretion of the vascular endothelial growth factor (VEGF) by keratinocytes and mast cells [92, 96]. An increased level of VEGF contributes to the dilatation of capillaries, erythema, and inflammatory edema characteristics of AD [97, 98]. Moreover, IL-9 has been shown to regulate Th1-mediated allergic contact dermatitis. Patients with positive patch tests to nickel have a higher level of allergen-specific IL-9 expression in skin, peripheral blood mononuclear cells (PBMCs). Also, IL-9 potentially mediates infiltration of eosinophils in the skins as its levels strongly correlate with the cell infiltration in the tissues. This demonstrates a potential pathogenic role of the cytokine IL-9 in ACD [99, 100].

9. IL-9 and autoimmunity

The etiology or trigger of autoimmune diseases is not well understood [101, 102]. However, there is a consensus that many factors, including genetic, environmental, and cytokine dysregulation are implicated in causing aberrant immune responses that drive tissue damage [102–104]. Many studies on divergent immune responses in autoimmunity have shown dysfunction of helper T cell subsets, which include Th1, Th17, and/or Treg cells [104, 105]. Studies in the last decade have identified IL-9-secreting Th9 cells as another T helper cell subset involved in immune responses [23, 106]. The IL-9 cytokine has become the focus of many autoimmune studies [107, 108]. Initial studies showed IL-9 to be a growth factor and a Th2 cytokine [13, 108]. More recently, IL-9 has been characterized as a lineage-specific cytokine for Th9 cells [109]. Thereafter, many immune cells involved in autoimmunity, such as Th17 and Treg cells, have demonstrated secretion of IL-9 [16, 110]. In EAE, a rodent model of MS, researchers identified Th9 and its signature cytokine, IL-9, in driving the disease process [111]. Its close association with Th17 and TGF-β has renewed interest in the role of IL-9 in the pathogenesis of autoimmune diseases [23]. In this section, we will examine the role of IL-9 in some autoimmune diseases such as multiple sclerosis (MS), systemic lupus erythematosus (SLE), inflammatory bowel diseases (IBD), rheumatoid arthritis (RA), and uveitis.

10. IL-9 and IL-17 dynamics in autoimmunity

The role of IL-9 in autoimmunity was illuminated when many studies reported that IL-9 and IL-17 are intricately related in driving the pathogenesis of diseases [111]. Human and animal studies revealed that Th17 cells secrete some amount of IL-9, in addition to other proinflammatory cytokines [112]. During the differentiation of naive T cells, TGF-β, a key driver of Th17 polarization, plays an important role in the differentiation of Th9 cells [23]. This was well elaborated in a study by Nowak *et al* in which *in vitro* polarization of MOG-specific Th17 cells was shown to generate IL-9-secreting Th9 [22, 113]. Secretion of IL-9 was further enhanced by the addition of IL-1β or IL-21 to the culture [113]. In addition, TGF-β and IL-6 induce Th17 cells that co-express IL-9 and IL-17 [22]. Studies have shown an increased frequency of memory CD4 cells that co-express IL-9 and IL-17 in patients with Type 1 diabetes [23].

On the other hand, IL-9 potentiates Th17 functions in an autocrine manner on Th17 cells [22, 110]. Th17 is a predominant helper T-cell subset that expresses IL-9 receptors (IL-9R) [22] . Through this receptor, IL-9 acts as an activator of Th17 cells [22]. IL-9 also synergizes with TGF-β to differentiate naive T cells into Th17 cells [110]. The presence of IL-9 in T cell cultures leads to the expansion of Th17 cells [110]. The importance of IL-9 in Th17 cell function is emphasized in IL-9R-deficient experimental autoimmune encephalomyelitis (EAE) model. Mice that lack IL-9 signaling showed decreased Th17 cells and defective migration of Th17 cells into the CNS [22, 114]. Neutralization of IL-9 led to attenuation of disease in EAE [22]. This unique relationship between IL-9 and Th17 provides the premise to examine the role of IL-9 in Th17-mediated autoimmune diseases.

11. Multiple sclerosis (MS)

Most autoimmune diseases like MS occur due to alteration of immune responses, which leads to tissue damage. The importance of IL-9 in MS has been enhanced through our understanding of the roles of IL-9-secreting T cells in EAE, an animal model of MS orchestrated by helper T cells [115]. Most studies revealed IL-9 plays a pathogenic role in EAE [22]. Th9 cells and Th17 cells were observed in the central nervous system (CNS) during EAE [115]. Blockade of IL-9 signaling in EAE resulted in contradictory conclusions. One study reported increased severity of disease in IL9Ra KO mice on a C57BL/6 background through a loss of Treg function and increased secretion of GM-CSF [116]. Other studies showed attenuation of disease and decreased Th17 cell infiltration into the CNS of SJL mice treated with IL-9 blocking antibody [22, 117]. This opposing view in disease outcome may be due to differences in the helper T cell composition and dysfunction driving the pathogenesis in the mouse strains. Also, IL-9 has been shown to increase chemokine CCL20, which enhances migration of Th17 into the CNS [22]. Accumulation and activation of mast cells during the Th17-IL9 immune response could explain the feedback loop [113]. Adoptive transfer of IL-9$^+$ Th9 into recipient mice resulted in EAE [118]. Th9-EAE model manifested a unique disease profile independent of Th1 and Th17 EAE models [118].

The role of IL-9 in MS patients is complex. A study by Roucco *et al* showed that IL-9 activates STAT1 and STAT 5, which are inhibitors of Th17 function [119]. IL-9 directly interfered with IL-17 expression in Th17 cells. Levels of IL-9 in the cerebrospinal fluid (CSF) of relapsing and remitting MS patients were inversely correlated with the disease pathogenesis and the disability indices [119]. These findings suggested the immunoregulatory role of IL-9 in MS. In another study, CSF of MS patients showed increased amounts of IL-9, and levels of IL-9 correlated well with IL-17 [120]. Therefore, more studies are needed to understand the functional role of IL-9 in MS.

12. Uveitis

Unlike other autoimmune diseases, uveitis is a heterogeneous disorder that results in inflammation of the eye [121]. In animal models of uveitis, adoptive transfer of *in vitro* polarized Th9 cells induced ocular inflammation [122, 123]. However, IL-9 was not detected in the eyes or lymph nodes of these mice [123]. Analysis of inflammatory cytokines in the vitreous humor of patients with uveitis detected increased levels of IL-9, among other proinflammatory cytokines [124]. However, the biological relevance of increased IL-9 in the study was not elaborated.

Another study examined the role of IL-9 in patients with Vogt-Koyanagi-Harada (VKH) disease. VKH is a systemic autoimmunity that manifests with bilateral panuveitis [125]. Patients with active disease had significantly higher levels of IL-9 in culture supernatants and higher IL-9 mRNA in PBMCs than did healthy controls and inactive patients [126]. The synergy of IL-9 and IL-17 was demonstrated in the study. The secretion of IL-17 by IL-9-treated PBMCs of active patients was significantly higher compared to the controls or inactive patients [126]. In a study that evaluated the serum of patients with Behcet's disease, another complex autoimmune disease with uveitis, serum IL-9 was neither elevated in disease state nor correlated with disease index [127]. More studies are needed to understand whether IL-9 signaling plays any immunological role in the eye.

13. Rheumatoid arthritis (RA)

The study of IL-9 in RA highlights its functional relationship with Tregs. In an antigen-induced animal model of arthritis, mice that lacked IL-9 had a chronic disease [128]. Treatment with rIL-9 resolved the joint inflammation, swelling, and tissue damage. The absence of IL-9 led to impaired suppressive functions of Treg cells [128]. Type 2 innate lymphoid cells (IL-C2) are documented to express IL-9 and have an anti-inflammatory function [128, 129]. These studies highlight the role of IL-9 in the resolution of inflammation in arthritis [130]. In human studies, IL-9-producing IL-C2 cells were also identified in the PBMCs of RA patients [130, 131]. In a study of treatment-induced remission of RA, synovial fluid of patients showed high levels of IL-9 [128].

14. Systemic lupus erythematosus (SLE)

Proinflammatory cytokines are generally believed to be involved in the pathogenesis of SLE. High levels of IL-9 mRNA and Th17 cells were seen in SLE patients compared with healthy controls (HC) [132, 133]. Dantas *et al,* evaluated the level of IL-9 in SLE and observed that patients with SLE had elevated IL-9 compared with levels in healthy individuals [134]. Further, IL-9$^+$ CD4 cells were more abundant in patients with SLE [132]. Serum IL-9 and mRNA of IL-9 were significantly elevated in SLE patients [132]. Also the elevated serum IL-9 and mRNA correlated with the SLE severity index [132, 135]. Animal studies corroborated these findings. Spleens and kidneys of lupus-prone mice showed high expression of IL-9 [136]. Neutralizing antibodies of IL-9 decreased kidney manifestation of SLE (lupus nephritis) and decreased anti-dsDNA antibody titers in these animal models [136].

15. Inflammatory bowel disease (IBD)

Aberrant adaptive immune response to the gut epithelial cells involving both CD4 and CD8 is implicated in the IBD [137]. These T cells are shown to express α4/β7 integrin, which binds to MAdcam1 on the gut epithelium [138, 139]. Gut T cells including cells that secrete IL-9 have been shown to express high levels of this integrin, and they propagate inflammation in the gut [140]. Gene expression studies have highlighted IR4 and GATA3 expression on immune cells that reside in the epithelial lining of the gut [141]. IRF4 is a transcription factor that drives the induction of Th9 immune responses in the gut [141]. Animal models of colitis confirms this finding of an abundance of the IL-9-producing T cells in the gut. These T-cells-producing IL-9 are involved in breaking the intestinal barrier [142]. In a DSS colitis model, anti-IL-9 blocking antibodies suppressed mucosal inflammation, and attenuation of disease was observed [142]. Adoptive transfer of IL-9-producing T cells into Rag2 knockout ($Rag2^{-/-}$ KO) mice also induced colitis [143]. Furthermore, IL-9 was found to directly modulate the expression of tight junction proteins, claudin and occludin in the animal model of colitis [144]. This indicates that IL-9 directly inhibited membrane integrity.

Immunological assessment of patients with inflammatory bowel disease (IBD) revealed high expression of IL-9 in the lamina propria [145]. In addition to other gut-residing T cells in IBD, CD4 cells had increased production of proinflammatory cytokines, including IL-9, which drive gut inflammation [145, 146]. Elevated levels of IL-1β and IL-9 were observed in the serum of IBD patients, and these correlated with disease prognosis [147]. Epithelial cells of UC also showed high expression of IL-9 receptor (IL-9R) [147, 148]. This receptor expression is most pronounced in patients with active disease [147]. *Ex vivo* IL-9 treatment of intestinal epithelial cells from UC patients showed increased proliferation of epithelial cells and pSTAT 5 expression [110].

Together, these findings highlight the role of IL-9 in IBD and colitis models. IL-9 could serve as a therapeutic target for IBD. Mice treated with GATA 3 DNAzyme showed it directly reduced IL-9 production and some Th2 cytokines to attenuate disease [149].

16. Type I diabetes

Studies by Vasanthakumar *et al* examined the role of IL-9 in patients with diabetes mellitus (DM) [150]. They observed that memory T cells from patients stimulated with Th17 polarizing conditions led to IL-9 production [150]. This shows that Th17 cells from DM patients have an increased ability to secrete IL-9 [23]. The study also identified TGF-β as the critical activator of IL-9 secretion [23]. TGF-β activity links Th17 and IL-9 secretion.

IL-9 appears to play both anti- and pro-inflammatory functions in autoimmunity. The functional heterogeneity of IL-9 may result from the unique cells or the microenvironment producing it. In RA, IL-9 exhibits anti-inflammatory function [128]. Studies have elaborated the anti-inflammatory function of IL-9 as it potentiates Treg-dependent immune tolerance to allografts [151]. In the gut, it is regarded as proinflammatory [142]. Some studies have shown that the expression of the activation marker CD96 on Th9 cells may explain the immunological status of the secreted IL-9 [152]. Researchers have reported that Th9 with high expression of CD96 showed a reduced ability to cause colitis compared with Th9 with low expression of CD96, which is associated with severe intestinal inflammation [152]. More studies must be done to identify the immunological heterogeneity of IL-9.

17. IL-9 as a therapeutic target

One principle of treatment of autoimmune diseases involves inhibition of mediators of inflammation. Drugs that target proinflammatory cytokines are extensively used in the treatment of autoimmune diseases [153]. Here we explore the use of IL-9 blockade as a therapeutic target in different disease conditions.

Medimmune LLC developed a humanized anti-IL-9 monoclonal antibody, MEDI-528 [154]. This humanized anti-IL-9 monoclonal antibody was indicated for use in allergen-induced asthma in adults [154]. Results from the clinical trial of Medimmune MEDI-528 showed no increased efficacy in improving respiratory functions and control of asthma compared to placebo [155]. Preclinical studies in mice showed the efficacy of blocking IL-9 in maintaining the airway [156]. Questions remain regarding why therapy directed at IL-9 failed to produce the desired response in humans. Heterogeneity of IL-9 sources and functions could explain the differences in airway response observed in this clinic trial.

18. Other potential IL-9 treatments

IL-9R inhibitor (rhIL-9-ETA) is a chimeric toxin targeting IL9 receptor [157]. These IL-9R inhibitors have efficacy in targeting malignant cells in non-hodgkin's lymphoma (NHL) and acute myeloid leukemia (AML) expressing IL9 and IL-9R [157]. However, the efficacy of this drug has not been tested in autoimmunity. Pfizer Inc. developed a JAK/STAT pathway inhibitor, CP-690550 [158]. It specifically targets and inhibits the activation of JAK 3 [158]. This treatment effectively prevents transplant rejection [158]. This drug could be beneficial in inhibiting IL-9 signaling, which depends on the JAK/STAT pathway. JAK inhibitors have been used in the treatment of RA and psoriasis [159]. UC patients that were treated with JAK inhibitors showed decreased Th9 cells [160].

BNZ 132-1-40 peptide, an antagonist of IL-2, IL-9, and IL-15 from Bioniz Therapeutics is undergoing safety and tolerability testing in patients with moderate to severe alopecia areata, an autoimmune disease of the skin that leads to hair loss [161]. However, no results from the clinical trial were available at the time of this review. Recently, FDA approved the use of BNZ-1 for the treatment of cutaneous T cell lymphoma (CTCL) [162]. These studies suggest BNZ-1 could be used to target IL-9 in diseases [163].

Other potential drug options include RDP58, which targets IRF4, a transcription factor involved in Th9 induction [164]. Interferon gamma (IFN-γ) has the ability to inhibit Th9 polarization through IL-27-dependent mechanisms [165]. Actimmune, an IFN-γ-based therapy by Horizon Therapeutics, is FDA-approved for the treatment of chronic granulomatous disease (CGD) [166]. The efficacy of inhibiting IL-9 by this drug could be tested in IL-9-related disorders.

The immune modulatory roles of IL-9 in health and diseases are important and provides a basis for exploring IL-9 as a therapeutic target. However, the divergent roles of IL-9 in promoting and inhibiting inflammation complicate definitive drug development. Some studies have highlighted the function of IL-9 in promoting immune tolerance. Future studies to understand cell-specific IL-9 regulation and function may resolve the conundrum of therapy development targeting IL-9. More studies in disease will broaden our knowledge about IL-9 function.

19. Conclusion

Significant progress has been made in our understanding of the functions of IL9 in health and diseases. For a long time, IL-9 was considered as a T cell growth factor, however, the identification of Th9 helper T cells has expanded our understanding on the roles IL-9 play in diseases. The pathogenic functions of IL-9 in autoimmunity and allergy suggest that IL-9 signaling can be targeted for therapy development. In this chapter, we focused on the function of IL-9 in different autoimmune diseases that include MS, SLE, RA, uveitis, and allergic conditions. We also highlighted IL-9-Th17 paradigm and its complexity in autoimmune diseases. Animal models of autoimmune diseases revealed contrasting roles of IL-9 and human studies are limited. Therefore, extensive animal and human research are necessary to elucidate the divergent immunological roles of IL-9. Such studies will be required for effective drug development that targets IL-9 signaling.

Acknowledgements

We will like to thank Kathy Kyler of the Office of the Vice President for Research and Mary Carter Ph.D. of the Writing center, University of Oklahoma Health Sciences center. Also, we will like to extend our gratitude to Dr. Jimmy Ballard and the staff of Microbiology and Immunology, OUHSC.

Conflict of interest

The authors declare no conflict of interests.

Author details

Ahmed Ummey Khalecha Bintha[1], Amani Souwelimatou Amadou[1], Mursalin Md Huzzatul[2] and Muhammad Fauziyya[3*]

1 Department of Microbiology and Immunology, University of Oklahoma Health Sciences Center, Oklahoma City, Oklahoma, USA

2 Department of Ophthalmology, University of Oklahoma Health Sciences Center, Oklahoma City, Oklahoma, USA

3 Department of Neurosurgery, University of Oklahoma Health Sciences Center, Oklahoma City, Oklahoma, USA

*Address all correspondence to: fauziyya-muhammad@ouhsc.edu

References

[1] Goswami, R. and M.H. Kaplan, *A brief history of IL-9*. J Immunol, 2011. 186(6): p. 3283-3288.

[2] Uyttenhove, C., R.J. Simpson, and J. Van Snick, *Functional and structural characterization of P40, a mouse glycoprotein with T-cell growth factor activity*. Proc Natl Acad Sci U S A, 1988. **85**(18): p. 6934-6938.

[3] Van Snick, J., et al., *Cloning and characterization of a cDNA for a new mouse T cell growth factor (P40)*. J Exp Med, 1989. **169**(1): p. 363-368.

[4] Hültner, L., et al., *Mast cell growth-enhancing activity (MEA) is structurally related and functionally identical to the novel mouse T cell growth factor P40/TCGFIII (interleukin 9)*. Eur J Immunol, 1990. **20**(6): p. 1413-6.

[5] Renauld, J.C., et al., *Human P40/IL-9. Expression in activated CD4+ T cells, genomic organization, and comparison with the mouse gene*. J Immunol, 1990. **144**(11): p. 4235-4241.

[6] Mock, B.A., et al., *IL9 maps to mouse chromosome 13 and human chromosome 5*. Immunogenetics, 1990. **31**(4): p. 265-70.

[7] Simpson, R.J., et al., *Complete amino acid sequence of a new murine T-cell growth factor P40*. Eur J Biochem, 1989. **183**(3): p. 715-722.

[8] Koh, B., et al., *A conserved enhancer regulates Il9 expression in multiple lineages*. Nat Commun, 2018. **9**(1): p. 4803.

[9] Kaplan, M.H., *The transcription factor network in Th9 cells*. Semin Immunopathol, 2017. **39**(1): p. 11-20.

[10] Staudt, V., et al., *Interferon-regulatory factor 4 is essential for the developmental program of T helper 9 cells*. Immunity, 2010. **33**(2): p. 192-202.

[11] Jash, A., et al., *Nuclear factor of activated T cells 1 (NFAT1)-induced permissive chromatin modification facilitates nuclear factor-κB (NF-κB)-mediated interleukin-9 (IL-9) transactivation*. J Biol Chem, 2012. **287**(19): p. 15445-15457.

[12] Gessner, A., H. Blum, and M. Röllinghoff, *Differential regulation of IL-9-expression after infection with Leishmania major in susceptible and resistant mice*. Immunobiology, 1993. **189**(5): p. 419-435.

[13] Else, K.J., L. Hültner, and R.K. Grencis, *Cellular immune responses to the murine nematode parasite Trichuris muris. II. Differential induction of TH-cell subsets in resistant versus susceptible mice*. Immunology, 1992. **75**(2): p. 232-237.

[14] Schmitt, E., et al., *TCGF III/P40 is produced by naive murine CD4+ T cells but is not a general T cell growth factor*. Eur J Immunol, 1989. **19**(11): p. 2167-2170.

[15] Temann, U.A., et al., *Expression of interleukin 9 in the lungs of transgenic mice causes airway inflammation, mast cell hyperplasia, and bronchial hyperresponsiveness*. J Exp Med, 1998. **188**(7): p. 1307-1320.

[16] Chang, H.C., et al., *The transcription factor PU.1 is required for the development of IL-9-producing T cells and allergic inflammation*. Nat Immunol, 2010. **11**(6): p. 527-534.

[17] Dardalhon, V., et al., *IL-4 inhibits TGF-beta-induced Foxp3+ T cells and, together with TGF-beta, generates IL-9+ IL-10+ Foxp3(−) effector T cells*. Nat Immunol, 2008. **9**(12): p. 1347-1355.

[18] Veldhoen, M., et al., *Transforming growth factor-beta 'reprograms' the differentiation of T helper 2 cells and promotes an interleukin 9-producing*

subset. Nat Immunol, 2008. **9**(12): p. 1341-1346.

[19] Li, M.O., Y.Y. Wan, and R.A. Flavell, *T cell-produced transforming growth factor-beta1 controls T cell tolerance and regulates Th1- and Th17-cell differentiation.* Immunity, 2007. **26**(5): p. 579-591.

[20] Schmitt, E., et al., *IL-9 production of naive CD4+ T cells depends on IL-2, is synergistically enhanced by a combination of TGF-beta and IL-4, and is inhibited by IFN-gamma.* J Immunol, 1994. **153**(9): p. 3989-3996.

[21] Tesmer, L.A., et al., *Th17 cells in human disease.* Immunol Rev, 2008. **223**: p. 87-113.

[22] Nowak, E.C., et al., *IL-9 as a mediator of Th17-driven inflammatory disease.* J Exp Med, 2009. **206**(8): p. 1653-1660.

[23] Beriou, G., et al., *TGF-beta induces IL-9 production from human Th17 cells.* J Immunol, 2010. **185**(1): p. 46-54.

[24] Putheti, P., et al., *Human CD4 memory T cells can become CD4+IL-9+ T cells.* PLoS One, 2010. **5**(1): p. e8706.

[25] Hültner, L., et al., *In activated mast cells, IL-1 up-regulates the production of several Th2-related cytokines including IL-9.* J Immunol, 2000. **164**(11): p. 5556-63.

[26] Stassen, M., et al., *IL-9 and IL-13 production by activated mast cells is strongly enhanced in the presence of lipopolysaccharide: NF-kappa B is decisively involved in the expression of IL-9.* J Immunol, 2001. **166**(7): p. 4391-8.

[27] Stassen, M., et al., *p38 MAP kinase drives the expression of mast cell-derived IL-9 via activation of the transcription factor GATA-1.* Mol Immunol, 2007. **44**(5): p. 926-33.

[28] Stassen, M., et al., *Murine bone marrow-derived mast cells as potent producers of IL-9: costimulatory function of IL-10 and kit ligand in the presence of IL-1.* J Immunol, 2000. **164**(11): p. 5549-55.

[29] Wiener, Z., A. Falus, and S. Toth, *IL-9 increases the expression of several cytokines in activated mast cells, while the IL-9-induced IL-9 production is inhibited in mast cells of histamine-free transgenic mice.* Cytokine, 2004. **26**(3): p. 122-130.

[30] Lauwerys, B.R., et al., *Cytokine production and killer activity of NK/T-NK cells derived with IL-2, IL-15, or the combination of IL-12 and IL-18.* J Immunol, 2000. **165**(4): p. 1847-53.

[31] Nagato, T., et al., *Expression of interleukin-9 in nasal natural killer/T-cell lymphoma cell lines and patients.* Clin Cancer Res, 2005. **11**(23): p. 8250-8257.

[32] Jones, T.G., et al., *Antigen-induced increases in pulmonary mast cell progenitor numbers depend on IL-9 and CD1d-restricted NKT cells.* J Immunol, 2009. **183**(8): p. 5251-5260.

[33] Gounni, A.S., et al., *IL-9 expression by human eosinophils: regulation by IL-1beta and TNF-alpha.* J Allergy Clin Immunol, 2000. **106**(3): p. 460-466.

[34] Sun, B., et al., *Characterization and allergic role of IL-33-induced neutrophil polarization.* Cell Mol Immunol, 2018. **15**(8): p. 782-793.

[35] Xiao, M., et al., *Osteoblasts support megakaryopoiesis through production of interleukin-9.* Blood, 2017. **129**(24): p. 3196-3209.

[36] Russell, S.M., et al., *Interaction of IL-2R beta and gamma c chains with Jak1 and Jak3: implications for XSCID and XCID.* Science, 1994. **266**(5187): p. 1042-1045.

[37] Demoulin, J.B. and J.C. Renauld, *Signalling by cytokines interacting with the interleukin-2 receptor gamma chain.* Cytokines Cell Mol Ther, 1998. **4**(4): p. 243-256.

[38] Renauld, J.C., et al., *Expression cloning of the murine and human interleukin 9 receptor cDNAs.* Proc Natl Acad Sci U S A, 1992. **89**(12): p. 5690-5694.

[39] Kimura, Y., et al., *Sharing of the IL-2 receptor gamma chain with the functional IL-9 receptor complex.* Int Immunol, 1995. **7**(1): p. 115-120.

[40] Bauer, J.H., et al., *Heteromerization of the gammac chain with the interleukin-9 receptor alpha subunit leads to STAT activation and prevention of apoptosis.* J Biol Chem, 1998. **273**(15): p. 9255-9260.

[41] Zhu, Y.X., et al., *Critical cytoplasmic domains of human interleukin-9 receptor alpha chain in interleukin-9-mediated cell proliferation and signal transduction.* J Biol Chem, 1997. **272**(34): p. 21334-21340.

[42] Malka, Y., et al., *Ligand-independent homomeric and heteromeric complexes between interleukin-2 or -9 receptor subunits and the gamma chain.* J Biol Chem, 2008. **283**(48): p. 33569-33577.

[43] Fujiwara, H., et al., *Homodimerization of the human interleukin 4 receptor alpha chain induces Cepsilon germline transcripts in B cells in the absence of the interleukin 2 receptor gamma chain.* Proc Natl Acad Sci U S A, 1997. **94**(11): p. 5866-5871.

[44] Ihle, J.N. and I.M. Kerr, *Jaks and Stats in signaling by the cytokine receptor superfamily.* Trends Genet, 1995. **11**(2): p. 69-74.

[45] Demoulin, J.B., et al., *A single tyrosine of the interleukin-9 (IL-9) receptor is required for STAT activation, antiapoptotic activity, and growth regulation by IL-9.* Mol Cell Biol, 1996. **16**(9): p. 4710-6.

[46] Levy, D.E. and J.E. Darnell, Jr., *Stats: transcriptional control and biological impact.* Nat Rev Mol Cell Biol, 2002. **3**(9): p. 651-662.

[47] Malik, S. and A. Awasthi, *Transcriptional Control of Th9 Cells: Role of Foxo1 in Interleukin-9 Induction.* Frontiers in Immunology, 2018. **9**: p. 995.

[48] Humblin, E., et al., *IRF8-dependent molecular complexes control the Th9 transcriptional program.* Nature Communications, 2017. **8**(1): p. 2085.

[49] Ngoc, L.P., et al., *Cytokines, allergy, and asthma.* Current opinion in allergy and clinical immunology, 2005. **5**(2): p. 161-166.

[50] Veldhoen, M., et al., *Transforming growth factor-β'reprograms' the differentiation of T helper 2 cells and promotes an interleukin 9–producing subset.* Nature immunology, 2008. **9**(12): p. 1341-1346.

[51] Gessner, A., H. Blum, and M. Röllinghoff, *Differential regulation of IL-9-expression after infection with Leishmania major in susceptible and resistant mice.* Immunobiology, 1993. **189**(5): p. 419-435.

[52] Dugas, B., et al., *Interleukin-9 potentiates the interleukin-4-induced immunoglobulin (IgG, IgM and IgE) production by normal human B lymphocytes.* European journal of immunology, 1993. **23**(7): p. 1687-1692.

[53] Takatsuka, S., et al., *IL-9 receptor signaling in memory B cells regulates humoral recall responses.* Nature immunology, 2018. **19**(9): p. 1025-1034.

[54] MURPHY, K. and C. WEAVER, *JANEWAY'S 9TH EDITION.*

[55] Petit-Frere, C., et al., *Interleukin-9 potentiates the interleukin-4-induced IgE and IgG1 release from murine B lymphocytes.* Immunology, 1993. **79**(1): p. 146.

[56] Dong, Q., et al., *IL-9 induces chemokine expression in lung epithelial cells and baseline airway eosinophilia in transgenic mice.* European journal of immunology, 1999. **29**(7): p. 2130-2139.

[57] Longphre, M., et al., *Allergen-induced IL-9 directly stimulates mucin transcription in respiratory epithelial cells.* The Journal of clinical investigation, 1999. **104**(10): p. 1375-1382.

[58] Sherkat, R., et al., *Innate lymphoid cells and cytokines of the novel subtypes of helper T cells in asthma.* Asia Pacific Allergy, 2014. **4**(4): p. 212-221.

[59] Hoppenot, D., et al., *Peripheral blood Th9 cells and eosinophil apoptosis in asthma patients.* Medicina, 2015. **51**(1): p. 10-17.

[60] Erpenbeck, V.J., et al., *Increased expression of interleukin-9 messenger RNA after segmental allergen challenge in allergic asthmatics.* Chest, 2003. **123**(3): p. 370S.

[61] Kim, M.S., et al., *Effects of interleukin-9 blockade on chronic airway inflammation in murine asthma models.* Allergy, asthma & immunology research, 2013. **5**(4): p. 197-206.

[62] Fawaz, L.M., et al., *Expression of IL-9 receptor α chain on human germinal center B cells modulates IgE secretion.* Journal of allergy and clinical immunology, 2007. **120**(5): p. 1208-1215.

[63] Gounni, A.S., et al., *IL-9-mediated induction of eotaxin1/CCL11 in human airway smooth muscle cells.* The Journal of Immunology, 2004. **173**(4): p. 2771-2779.

[64] Levitt, R.C., et al., *IL-9 pathway in asthma: new therapeutic targets for allergic inflammatory disorders.* Journal of Allergy and Clinical Immunology, 1999. **103**(5): p. S485-S491.

[65] Temann, U.-A., et al., *Expression of interleukin 9 in the lungs of transgenic mice causes airway inflammation, mast cell hyperplasia, and bronchial hyperresponsiveness.* The Journal of experimental medicine, 1998. **188**(7): p. 1307-1320.

[66] Wiener, Z., A. Falus, and S. Toth, *IL-9 increases the expression of several cytokines in activated mast cells, while the IL-9-induced IL-9 production is inhibited in mast cells of histamine-free transgenic mice.* Cytokine, 2004. **26**(3): p. 122-130.

[67] Matsuzawa, S., et al., *IL-9 enhances the growth of human mast cell progenitors under stimulation with stem cell factor.* The Journal of Immunology, 2003. **170**(7): p. 3461-3467.

[68] Temann, U.-A., et al., *IL9 leads to airway inflammation by inducing IL13 expression in airway epithelial cells.* International immunology, 2007. **19**(1): p. 1-10.

[69] McLane, M.P., et al., *Interleukin-9 promotes allergen-induced eosinophilic inflammation and airway hyperresponsiveness in transgenic mice.* American journal of respiratory cell and molecular biology, 1998. **19**(5): p. 713-720.

[70] Ciprandi, G., *Serum interleukin 9 in allergic rhinitis.* Annals of Allergy, Asthma & Immunology, 2010. **104**(2): p. 180-181.

[71] Gu, Z.W., Y.X. Wang, and Z.W. Cao, *Neutralization of interleukin-9 ameliorates symptoms of allergic rhinitis by reducing Th2, Th9, and Th17 responses and increasing the Treg response in a murine model.* Oncotarget, 2017. **8**(9): p. 14314.

[72] Osterfeld, H., et al., *Differential roles for the IL-9/IL-9 receptor alpha-chain pathway in systemic and oral antigen-induced anaphylaxis.* J Allergy Clin Immunol, 2010. **125**(2): p. 469-476.e2.

[73] Nakajima-Adachi, H., et al., *Critical role of intestinal interleukin-4 modulating regulatory T cells for desensitization, tolerance, and inflammation of food allergy.* PLoS One, 2017. **12**(2): p. e0172795.

[74] Burton, O.T., et al., *Direct effects of IL-4 on mast cells drive their intestinal expansion and increase susceptibility to anaphylaxis in a murine model of food allergy.* Mucosal Immunol, 2013. **6**(4): p. 740-750.

[75] Shik, D., et al., *IL-9-producing cells in the development of IgE-mediated food allergy.* Semin Immunopathol, 2017. **39**(1): p. 69-77.

[76] El Ansari, Y.S., C. Kanagaratham, and H.C. Oettgen, *Mast Cells as Regulators of Adaptive Immune Responses in Food Allergy.* Yale J Biol Med, 2020. **93**(5): p. 711-718.

[77] Ventura, M., et al., *Intestinal permeability in patients with adverse reactions to food.* Digestive and liver disease, 2006. **38**(10): p. 732-736.

[78] Xie, J., et al., *Elevated antigen-driven IL-9 responses are prominent in peanut allergic humans.* PloS one, 2012. **7**(10): p. e45377.

[79] Brough, H.A., et al., *IL-9 is a key component of memory TH cell peanut-specific responses from children with peanut allergy.* Journal of allergy and clinical immunology, 2014. **134**(6): p. 1329-1338. e10.

[80] Chen, C.Y., et al., *Induction of Interleukin-9-Producing Mucosal Mast Cells Promotes Susceptibility to IgE-Mediated Experimental Food Allergy.* Immunity, 2015. **43**(4): p. 788-802.

[81] Tomar, S., et al., *IL-4–BATF signaling directly modulates IL-9 producing mucosal mast cell (MMC9) function in experimental food allergy.* Journal of Allergy and Clinical Immunology, 2021. **147**(1): p. 280-295.

[82] Forbes, E.E., et al., *IL-9- and mast cell-mediated intestinal permeability predisposes to oral antigen hypersensitivity.* J Exp Med, 2008. **205**(4): p. 897-913.

[83] Osterfeld, H., et al., *Differential roles for the IL-9/IL-9 receptor α-chain pathway in systemic and oral antigen–induced anaphylaxis.* Journal of allergy and clinical immunology, 2010. **125**(2): p. 469-476. e2.

[84] Ahrens, R., et al., *Intestinal mast cell levels control severity of oral antigen-induced anaphylaxis in mice.* The American journal of pathology, 2012. **180**(4): p. 1535-1546.

[85] Chen, C.-Y., et al., *Induction of interleukin-9-producing mucosal mast cells promotes susceptibility to IgE-mediated experimental food allergy.* Immunity, 2015. **43**(4): p. 788-802.

[86] Tomar, S., et al., *IL-4-BATF signaling directly modulates IL-9 producing mucosal mast cell (MMC9) function in experimental food allergy.* Journal of Allergy and Clinical Immunology, 2020.

[87] Steenwinckel, V., et al., *IL-9 promotes IL-13-dependent paneth cell hyperplasia and up-regulation of innate immunity mediators in intestinal mucosa.* The journal of immunology, 2009. **182**(8): p. 4737-4743.

[88] Forbes, E.E., et al., *IL-9–and mast cell–mediated intestinal permeability predisposes to oral antigen hypersensitivity.* The Journal of experimental medicine, 2008. **205**(4): p. 897-913.

[89] Ma, L., et al., *Possible pathogenic role of T helper type 9 cells and interleukin*

(IL)-9 in atopic dermatitis. Clin Exp Immunol, 2014. **175**(1): p. 25-31.

[90] Klonowska, J., et al., *New Cytokines in the Pathogenesis of Atopic Dermatitis-New Therapeutic Targets.* Int J Mol Sci, 2018. **19**(10).

[91] Ciprandi, G., et al., *Serum interleukin-9 levels are associated with clinical severity in children with atopic dermatitis.* Pediatric dermatology, 2013. **30**(2): p. 222-225.

[92] Ma, L., et al., *Possible pathogenic role of T helper type 9 cells and interleukin (IL)-9 in atopic dermatitis.* Clinical & Experimental Immunology, 2014. **175**(1): p. 25-31.

[93] Namkung, J.-H., et al., *An association between IL-9 and IL-9 receptor gene polymorphisms and atopic dermatitis in a Korean population.* Journal of dermatological science, 2011. **62**(1): p. 16-21.

[94] Stockinger, B.B., et al., *Interleukin 9 fate reporter reveals induction of innate IL-9 response in lung* inflammation. 2011.

[95] Brunner, P.M., E. Guttman-Yassky, and D.Y. Leung, *The immunology of atopic dermatitis and its reversibility with broad-spectrum and targeted therapies.* Journal of Allergy and Clinical Immunology, 2017. **139**(4): p. S65-S76.

[96] Sismanopoulos, N., et al., *IL-9 induces VEGF secretion from human mast cells and IL-9/IL-9 receptor genes are overexpressed in atopic dermatitis.* PLoS One, 2012. 7(3): p. e33271.

[97] Zhang, Y., H. Matsuo, and E. Morita, *Increased production of vascular endothelial growth factor in the lesions of atopic dermatitis.* Archives of dermatological research, 2006. **297**(9): p. 425.

[98] Chen, L., et al., *The progression of inflammation parallels the dermal angiogenesis in a keratin 14 IL-4-transgenic model of atopic dermatitis.* Microcirculation, 2008. **15**(1): p. 49-64.

[99] Liu, J., et al., *IL-9 regulates allergen-specific Th1 responses in allergic contact dermatitis.* Journal of Investigative Dermatology, 2014. **134**(7): p. 1903-1911.

[100] Louahed, J., et al., *Interleukin 9 promotes influx and local maturation of eosinophils.* Blood, The Journal of the American Society of Hematology, 2001. **97**(4): p. 1035-1042.

[101] Sarvetnick, N., *Etiology of autoimmunity.* Immunol Res, 2000. **21**(2-3): p. 357-362.

[102] Jörg, S., et al., *Environmental factors in autoimmune diseases and their role in multiple sclerosis.* Cell Mol Life Sci, 2016. **73**(24): p. 4611-4622.

[103] Smith, D.A. and D.R. Germolec, *Introduction to immunology and autoimmunity.* Environ Health Perspect, 1999. **107 Suppl 5**(Suppl 5): p. 661-5.

[104] Rosenblum, M.D., K.A. Remedios, and A.K. Abbas, *Mechanisms of human autoimmunity.* J Clin Invest, 2015. **125**(6): p. 2228-2233.

[105] Kaur, G., K. Mohindra, and S. Singla, *Autoimmunity-Basics and link with periodontal disease.* Autoimmun Rev, 2017. **16**(1): p. 64-71.

[106] Li, J., et al., *IL-9 and Th9 cells in health and diseases-From tolerance to immunopathology.* Cytokine Growth Factor Rev, 2017. **37**: p. 47-55.

[107] Pan, L.L., et al., *IL-9-producing Th9 cells may participate in pathogenesis of Takayasu's arteritis.* Clin Rheumatol, 2016. **35**(12): p. 3031-3036.

[108] Deng, Y., et al., *Th9 cells and IL-9 in autoimmune disorders: Pathogenesis and*

therapeutic potentials. Hum Immunol, 2017. **78**(2): p. 120-128.

[109] Weigmann, B. and M.F. Neurath, *Th9 cells in inflammatory bowel diseases.* Semin Immunopathol, 2017. **39**(1): p. 89-95.

[110] Elyaman, W., et al., *IL-9 induces differentiation of TH17 cells and enhances function of FoxP3+ natural regulatory T cells.* Proc Natl Acad Sci U S A, 2009. **106**(31): p. 12885-12890.

[111] Malik, S., V. Dardalhon, and A. Awasthi, *Characterization of Th9 Cells in the Development of EAE and IBD.* Methods Mol Biol, 2017. **1585**: p. 201-216.

[112] Noelle, R.J. and E.C. Nowak, *Cellular sources and immune functions of interleukin-9.* Nat Rev Immunol, 2010. **10**(10): p. 683-687.

[113] Nowak, E.C. and R.J. Noelle, *Interleukin-9 as a T helper type 17 cytokine.* Immunology, 2010. **131**(2): p. 169-173.

[114] Zhou, Y., et al., *IL-9 promotes Th17 cell migration into the central nervous system via CC chemokine ligand-20 produced by astrocytes.* J Immunol, 2011. **186**(7): p. 4415-4421.

[115] Elyaman, W. and S.J. Khoury, *Th9 cells in the pathogenesis of EAE and multiple sclerosis.* Semin Immunopathol, 2017. **39**(1): p. 79-87.

[116] Yoshimura, S., et al., *IL-9 Controls Central Nervous System Autoimmunity by Suppressing GM-CSF Production.* J Immunol, 2020. **204**(3): p. 531-539.

[117] Li, H., et al., *IL-9 is important for T-cell activation and differentiation in autoimmune inflammation of the central nervous system.* Eur J Immunol, 2011. **41**(8): p. 2197-2206.

[118] Jäger, A., et al., *Th1, Th17, and Th9 effector cells induce experimental autoimmune encephalomyelitis with different pathological phenotypes.* J Immunol, 2009. **183**(11): p. 7169-7177.

[119] Ruocco, G., et al., *T helper 9 cells induced by plasmacytoid dendritic cells regulate interleukin-17 in multiple sclerosis.* Clin Sci (Lond), 2015. **129**(4): p. 291-303.

[120] Khaibullin, T., et al., *Elevated Levels of Proinflammatory Cytokines in Cerebrospinal Fluid of Multiple Sclerosis Patients.* Front Immunol, 2017. **8**: p. 531.

[121] Squires, H., et al., *A systematic review and economic evaluation of adalimumab and dexamethasone for treating non-infectious intermediate uveitis, posterior uveitis or panuveitis in adults.* Health Technol Assess, 2017. **21**(68): p. 1-170.

[122] Kaplan, M.H., *Th9 cells: differentiation and disease.* Immunol Rev, 2013. **252**(1): p. 104-115.

[123] Horai, R. and R.R. Caspi, *Cytokines in autoimmune uveitis.* J Interferon Cytokine Res, 2011. **31**(10): p. 733-744.

[124] Fukunaga, H., et al., *Analysis of inflammatory mediators in the vitreous humor of eyes with pan-uveitis according to aetiological classification.* Sci Rep, 2020. **10**(1): p. 2783.

[125] Baltmr, A., S. Lightman, and O. Tomkins-Netzer, *Vogt-Koyanagi-Harada syndrome - current perspectives.* Clin Ophthalmol, 2016. **10**: p. 2345-2361.

[126] Peng, Z., et al., *Expression and role of interleukin-9 in Vogt-Koyanagi-Harada disease.* Mol Vis, 2017. **23**: p. 538-547.

[127] Nouri-Vaskeh, M., et al., *Lack of association between serum IL-9 levels and Behçet's disease.* Immunol Lett, 2019. **211**: p. 23-27.

[128] Rauber, S., et al., *Resolution of inflammation by interleukin-9-producing type 2 innate lymphoid cells.* Nat Med, 2017. **23**(8): p. 938-944.

[129] Karagiannis, F. and C. Wilhelm, *More Is Less: IL-9 in the Resolution of Inflammation.* Immunity, 2017. **47**(3): p. 403-405.

[130] Wu, X., *Innate Lymphocytes in Inflammatory Arthritis.* Front Immunol, 2020. **11**: p. 565275.

[131] Hughes-Austin, J.M., et al., *Multiple cytokines and chemokines are associated with rheumatoid arthritis-related autoimmunity in first-degree relatives without rheumatoid arthritis: Studies of the Aetiology of Rheumatoid Arthritis (SERA).* Ann Rheum Dis, 2013. **72**(6): p. 901-907.

[132] Ouyang, H., et al., *Increased interleukin-9 and CD4+IL-9+ T cells in patients with systemic lupus erythematosus.* Mol Med Rep, 2013. **7**(3): p. 1031-1037.

[133] Leng, R.X., et al., *Potential roles of IL-9 in the pathogenesis of systemic lupus erythematosus.* Am J Clin Exp Immunol, 2012. **1**(1): p. 28-32.

[134] Dantas, A.T., et al., *Increased Serum Interleukin-9 Levels in Rheumatoid Arthritis and Systemic Lupus Erythematosus: Pathogenic Role or Just an Epiphenomenon?* Dis Markers, 2015. **2015**: p. 519638.

[135] Ouyang, H., et al., *[Abnormality and significance of interleukin-9 and CD4(+)interleukin-9(+) T-cells in peripheral blood of patients with systemic lupus erythematosus].* Zhonghua Yi Xue Za Zhi, 2013. **93**(2): p. 99-103.

[136] Yang, J., et al., *Interleukin-9 Is Associated with Elevated Anti-Double-Stranded DNA Antibodies in Lupus-Prone Mice.* Mol Med, 2015. **21**(1): p. 364-370.

[137] Funderburg, N.T., et al., *Circulating CD4(+) and CD8(+) T cells are activated in inflammatory bowel disease and are associated with plasma markers of inflammation.* Immunology, 2013. **140**(1): p. 87-97.

[138] Wittner, M., et al., *Comparison of the integrin α4β7 expression pattern of memory T cell subsets in HIV infection and ulcerative colitis.* PLoS One, 2019. **14**(7): p. e0220008.

[139] Kurmaeva, E., et al., *T cell-associated α4β7 but not α4β1 integrin is required for the induction and perpetuation of chronic colitis.* Mucosal Immunol, 2014. **7**(6): p. 1354-1365.

[140] Jovani, M. and S. Danese, *Vedolizumab for the treatment of IBD: a selective therapeutic approach targeting pathogenic a4b7 cells.* Curr Drug Targets, 2013. **14**(12): p. 1433-1443.

[141] Malik, S. and A. Awasthi, *Transcriptional Control of Th9 Cells: Role of Foxo1 in Interleukin-9 Induction.* Front Immunol, 2018. **9**: p. 995.

[142] Yuan, A., et al., *IL-9 antibody injection suppresses the inflammation in colitis mice.* Biochem Biophys Res Commun, 2015. **468**(4): p. 921-926.

[143] de Heusch, M., et al., *IL-9 exerts biological function on antigen-experienced murine T cells and exacerbates colitis induced by adoptive transfer.* Eur J Immunol, 2020. **50**(7): p. 1034-1043.

[144] Gerlach, K., et al., *IL-9 regulates intestinal barrier function in experimental T cell-mediated colitis.* Tissue Barriers, 2015. **3**(1-2): p. e983777.

[145] Hufford, M.M. and M.H. Kaplan, *A gut reaction to IL-9.* Nat Immunol, 2014. **15**(7): p. 599-600.

[146] Matusiewicz, M., et al., *Systemic interleukin-9 in inflammatory bowel disease: Association with mucosal healing in ulcerative colitis.* World

J Gastroenterol, 2017. **23**(22): p. 4039-4046.

[147] Gerlach, K., et al., *TH9 cells that express the transcription factor PU.1 drive T cell-mediated colitis via IL-9 receptor signaling in intestinal epithelial cells.* Nat Immunol, 2014. **15**(7): p. 676-686.

[148] Vyas, S.P. and R. Goswami, *A Decade of Th9 Cells: Role of Th9 Cells in Inflammatory Bowel Disease.* Front Immunol, 2018. **9**: p. 1139.

[149] Popp, V., et al., *Rectal Delivery of a DNAzyme That Specifically Blocks the Transcription Factor GATA3 and Reduces Colitis in Mice.* Gastroenterology, 2017. **152**(1): p. 176-192.e5.

[150] Vasanthakumar, R., et al., *Serum IL-9, IL-17, and TGF-β levels in subjects with diabetic kidney disease (CURES-134).* Cytokine, 2015. **72**(1): p. 109-112.

[151] Burrell, B.E., et al., *Regulatory T cell induction, migration, and function in transplantation.* J Immunol, 2012. **189**(10): p. 4705-4711.

[152] Stanko, K., et al., *CD96 expression determines the inflammatory potential of IL-9-producing Th9 cells.* Proc Natl Acad Sci U S A, 2018. **115**(13): p. E2940-e2949.

[153] Willrich, M.A., D.L. Murray, and M.R. Snyder, *Tumor necrosis factor inhibitors: clinical utility in autoimmune diseases.* Transl Res, 2015. **165**(2): p. 270-282.

[154] Antoniu, S.A., *MEDI-528, an anti-IL-9 humanized antibody for the treatment of asthma.* Curr Opin Mol Ther, 2010. **12**(2): p. 233-239.

[155] White, B., et al., *Two first-in-human, open-label, phase I dose-escalation safety trials of MEDI-528, a monoclonal antibody against interleukin-9, in healthy adult volunteers.* Clin Ther, 2009. **31**(4): p. 728-740.

[156] Kim, M.S., et al., *Effects of interleukin-9 blockade on chronic airway inflammation in murine asthma models.* Allergy Asthma Immunol Res, 2013. **5**(4): p. 197-206.

[157] Klimka, A., et al., *A deletion mutant of Pseudomonas exotoxin-A fused to recombinant human interleukin-9 (rhIL-9-ETA') shows specific cytotoxicity against IL-9-receptor-expressing cell lines.* Cytokines Mol Ther, 1996. **2**(3): p. 139-146.

[158] Flanagan, M.E., et al., *Discovery of CP-690,550: a potent and selective Janus kinase (JAK) inhibitor for the treatment of autoimmune diseases and organ transplant rejection.* J Med Chem, 2010. **53**(24): p. 8468-8484.

[159] Kvist-Hansen, A., P.R. Hansen, and L. Skov, *Systemic Treatment of Psoriasis with JAK Inhibitors: A Review.* Dermatol Ther (Heidelb), 2020. **10**(1): p. 29-42.

[160] Imam, T., et al., *Effector T Helper Cell Subsets in Inflammatory Bowel Diseases.* Front Immunol, 2018. **9**: p. 1212.

[161] NCT03532958, P.T.o.B.-i.P.W.M.t.S.A.A., 2019.

[162] *A Dose-Ranging Study of IV BNZ-1 in LGL Leukemia or Refractory CTCL-NCT03239392.*

[163] Wang, T.T., et al., *IL-2 and IL-15 blockade by BNZ-1, an inhibitor of selective γ-chain cytokines, decreases leukemic T-cell viability.* Leukemia, 2019. **33**(5): p. 1243-1255.

[164] Li, J., et al., *Toll-like receptors as therapeutic targets for autoimmune connective tissue diseases.* Pharmacology & Therapeutics, 2013. **138**(3): p. 441-451.

[165] Murugaiyan, G., et al., *IFN-γ limits Th9-mediated autoimmune inflammation through dendritic cell modulation of IL-27.* J Immunol, 2012. **189**(11): p. 5277-5283.

[166] Green, D.S., et al., *Production of a cellular product consisting of monocytes stimulated with Sylatron® (Peginterferon alfa-2b) and Actimmune® (Interferon gamma-1b) for human use.* Journal of Translational Medicine, 2019. **17**(1): p. 82.

Chapter 5

From Pregnancy Loss to COVID 19 Cytokine Storm: A Matter of Inflammation and Coagulation

Fortunato Vesce

'Dosis facit venenum!'
Paracelso

Abstract

Large scientific evidence achieved during the second half of the past century points to a leading role of inflammation in the pathogenic mechanism of the main pregnancy complications, such as abortion, pregnancy loss, premature delivery, infection, fetal encephalopathy, enterocolitis, pulmonary hyaline membrane diseases and death. Thinking about pregnancy inflammation, one must refer today to the umbalance of the normal mediators of organic functions: cytokins, peptides, nucleosides, prostanoids. Indeed, according to the order and quantity of their release, they are involved either in physiology or in pathology of pregnancy. At this regard, it has been shown that Th1-type immunity is incompatible with successful pregnancy. Regulation of the mediators of maternal functions is largely under fetal genetic control. Assessment of the fetal role derives from studies showing an umbalance of cytokines and plasminogen activator system, an increase of endothelin, a downregulation of adenosine receptors, in the fetal compartment, in aneuploid pregnancies. The resulting functional deviations deal with inflammation, imfection, coagulation, impaired utero-placental perfusion, possibly leading to fetal demise and ominus maternal complications. SARS-COV-2 infection, on the other hand, is characterized by a similar umbalance of the inflammatory mediators, leading to hyperactivation of a type-1 lymphobyte T-helper response, which ends in a possibly fatal cytokine storm syndrome. While SARS-COV-2 infection recognizes a viral etiology, the cause of pregnancy inflammation must be recognized in the inability of the fetus to control the maternal immune response. Therefore, the preventive measures are quite different, although both benefit of a similar anti-inflammatory, antibiotic and anti-coagulant therapy.

Keywords: pregnancy inflammation, abortion, FIRS, SARS-COV-2, IL-6, viral pneumonia, cytokine storm

1. Introduction

Inflammation was defined in ancient times as *'rubor, calor, tumor, dolor and functio laesa'*: redness, heat, swelling, pain and functional impairment. However, in the large majority of the cases, this pathologic process starts before the onset of the clinical signs and symptoms, as a result of an unbalanced release of the mediators of tissue functions, among which cytokines and prostanoids. Such a release can be triggered by physical, chemical, metabolic, endocrine, infectious as well as mechanical events. Nevertheless, many normal functions are regulated by the same mediators that in other circumstances give rise to inflammation. Among physiological functions ovulation, menstruation, implantation of the product of conception, delivery, healing of the placental site and involution of the puerperal uterus are included. For instance, as regards the onset of parturition, we have shown that receptor ligands for the inflammatory peptide N-formyl-methionyl-leucyl-phenylalanine (fMLP) are present in amniotic fluid. Their levels do not vary during gestation, while they are significantly increased by labour, along with the expression of fMLP receptor in amnion tissue, thus indicating a modulation of the fMLP system by the events of physiological labour, and/or viceversa [1–3]. A detailed knowledge of cytokine and prostanoids involved in the regulation of normal pregnancy is needed to better understand the role of inflammatory mediators in the pathogenic mechanism of gestational complications.

At this regard it must first be considered that the trophoblast itself, i.e. the peripheral part of the product of conception, regulates implantation and placentation. These are a consequence of membrane ligands and receptors, hormone and local factor release by fetal and maternal tissues. There are two kinds of trophoblast, villous and extravillous, the first devoted to fetal-maternal nutrients and gas exchanges, and the second to adhesion of the placenta to the uterine wall and to the modulation of uterine arteries. Indeed, invasion of the uterine spiral arteries by extravillous throphoblast occurs, aimed at progressively increasing the perfusion of the intervillous space. A derangement of this stuctural and functional process leads to different types of complications, including pregnancy loss and maternal life-threatening disease.

In the past it was believed that these vascular changes occurred within the first half of pregnancy, but today it must be admitted that they last until the moment of delivery. From this point of view, pregnancy must be considered an endocrine mediated vascular phenomenon, regulated by cytokines mainly derived from the extra-villous trophobalst itself [4].

Trophoblast regulation is needed because the rupture of the spiral arterioles, with consequent dripping of the maternal blood in which the nutrition villi are immersed, in any other tissue except the uterine wall would trigger an inflammatory reaction aimed at coagulating the blood to stop hemorrhage. How is it possible that this defense mechanism is not activated in physiological pregnancy? As it will be explained, the reason is that extravillous trophoblast itself has the task of transforming the natural maternal TH-1 inflammatory reaction into an anti-inflammatory condition of the TH-2 type [5]. The lack of this transformation, in fact, leads to pregnancy loss and to other complications, such as premature birth, gestosis, fetal growth restriction and related postnatal syndromes [5]. For instance, as regards fetal growth, peripheral mononuclear cells stimulated with trophoblast antigens [6] as well as with mitogen [7] in pregnant women with fetal growth restriction produce higher levels of the pro-inflammatory cytokines IFNγ, TNFα, IL-8, IL-12, IL-18, IL-23 and lower anti-inflammatory cytokines IL-4, IL-10, IL-13 compared with pregnant women with normal fetal growth.

2. The misunderstood concept of 'maternal tolerance'

Although the scientific literature of the last few decades already contains evidence of the leading role of inflammatory cytokines in the mechanisms of pregnancy complications, it has not yet been completely understood by mainstream medicine. As a consequence the protective role of glucocorticoids administration is neglected or totally ignored.

Moreover, it has not been understood that physiological pregnancy does not compromise the general immunity of the maternal organism. At this regard, attempts to credit the hypothesis of a reduced maternal immune response are frequently made. A theory of maternal-fetal tolerance proposed that a 'temporary state of maternal immunosuppression' is vital to allow successful implantation and development of the product of conception [8, 9].

Subsequently, a tightly regulated balance between inflammatory and tolerogenic states during the "immune chronology" of normal pregnancy has been affirmed [10–14]. It is claimed that a pro-inflammatory environment predominates during early trophoblast invasion and at parturition, while it turns to anti-inflammatory during the second and third trimesters to allow fetal growth [15].

All true! What needs to be understood is the real nature of this changes. Indeed, it must be clear that it is absolutely wrong to speak of inflammation when a physiological function such as implantation or childbirth is triggered by the same mediators that, in pathologic conditions, would cause inflammation. Their derangement may trigger inflammation: *dosis facit venenum*, as Paracelso stated! It is also reported that dysregulation of immune cells at the level of maternal decidua is implicated in severe complications, including recurrent miscarriage, pre-eclampsia, fetal growth restriction, chorioamnionitis, and preterm birth [16–23].

It is therefore very important to understand the nature of the so called 'maternal tolerance' towards the product of conception, in order to avoid dangerous conclusions. Indeed, it is wrongly claimed that a corollary of this maternal tolerance implies an increased susceptibility to infection during pregnancy. In turn, this misinterpretation has generated the belief that pregnancy carries a high risk of severe flu syndrome which must be prevented by vaccination!

In order to better understand the terms of this matter, it must be briefly recalled that normally the changes in maternal immune system occur only at the utero-placental level, upon the direct action of the trophoblast. The features of these changes have nothing in common with the immune response to infections. As for the entire maternal organism, the integrity of the immune system is perfectly maintained. Both branches of immunity, that is, the humoral and the cellular, are fully operational during normal pregnancy. A clear example of the integrity of humoral immunity is represented, for instance, by maternal-fetal isoimmunization, i.e. Rh disease. In this pathologic condition the maternal immune system activates the production of antibodies against Rh positive fetal red blood cells leading to anemia, erythroblastosis and possibly to fetal death.

In the lack of a reliable demonstration of an increased incidence of flu and other infectious diseases during pregnancy, one may believe that, although active in Rh disease, the production of antibodies against viruses and bacteria is hindered. On the contrary, in fifty years of my personal clinical experience, I have never detected symptoms or signs of a reduced or ineffective maternal immune reactivity against infection, nor an increased incidence of infectious diseases in normal pregnancy. Moreover, looking at the scientific literature my conviction has been largely confirmed. Hundreds of articles including thousand of patients have been recently examined as far as flu is concerned. Contrary to the opinion of an increased risk and serious complications accredited by World Health Organization (WHO), a significantly lower risk of admission to Intensive Care Unit was registered for pregnant women. Moreover, no significant difference between pregnant and non-pregnant patients was registered, as regards mechanical ventilatory support. Pregnancy did not carry a greater likelihood of maternal death or other severe outcomes compared to either the general population or non-pregnant women of reproductive age. The only difference between pregnant and non-pregnant was a higher risk of hospitalization in the first, that the Authors correctly ascribed to a better care for motherhood [24].

Surprisingly, however, the above data have been interpreted in the opposite meaning by others, which also reported a disproportionaly high mortality rate in the flu pandemic of 1918 [25]. However, that was before the era of antibiotics, anticoagulants and anti-inflammatory drugs (the safety, preventive and therapeutic power of which is unfortunately still poorly understood today!).

3. How do infections affect pregnancy?

A further aspect of the matter is the influence of infections on the outcome of pregnancy. Indeed, an increased incidence of adverse events following viral infection would speak in favor of a preventive vaccination aimed at protecting pregnancy and the newborn future life.

It has been reported that infectious agents are potentially involved in about 40% of spontaneous abortion [26–28]. On the contrary, recent research failed to show an increased incidence of several infections in spontaneous abortion. The free mother–to-child transmission of the three oncogenic viruses, BKPyV, JCPyV and SV40 has been shown by detecting DNA sequences and specific IgG antibodies in mothers and their offspring in normal pregnancy [29]. The incidence of Human Papilloma Virus (HPV) infection is not increased in spontaneous abortion, and the overall prevalence of serum anti-HPV16 IgG antibodies was found to be 30% in patients with normal pregnancy and 37.5% in those with spontaneous abortion ($p > 0.05$), thus indicating a normal, or even better humoral immunity in the latter [30].

Rubella virus, varicella-zoster, human immunodeficiency virus, adenovirus, cytomegalovirus, herpes simplex virus, human parvovirus, Epstein–Barr virus, enterovirus and respiratory syncitial virus have all been found in amniotic fluid samples, but their mere presence has never been associated with negative human pregnancy outcome [31–34]. Based on the above evidences, it should appear that the gestational setting of human female immune system is towards a better protection against infectious diseases compared to non-pregnant.

Interestingly, however, it has been shown that viral experimental infection of pregnant mice predisposed to the effects of bacterial endotoxin [35]: it is an observation of extraordinary importance to better understand the pathogenesis of bacterial infections. Indeed, it explains that the bacteria normally present in their saprophytic state can become pathogenic as a consequence of a previously produced inflammation.

Accordingly, the onset of human preterm labor is preceded by a systemic fetal inflammation, before the appearance of clinical signs of maternal or fetal infection [36, 37]. It has been stated that an amniotic concentration of IL6 above 11 ng/ml is related to, and defines, the Systemic Fetal Inflammatory Response Syndrome, that is followed by premature birth and by utero-placental infection, with all the postnatal sequelae, including encephalopathy, enterocolitis and the pulmonary hyaline membrane disease.

Of particular importance is to consider what could be the origin of fetal inflammation, in the absence of maternal chronic inflammatory disease, pathogens and clinical signs of infection. Indeed, in primates the events leading to premature delivery seem to progress from experimental intrauterine infection to pro-inflammatory cytokine activation and prostaglandin production, thus triggering myometrial contractions [38–40]. In humans, instead, also the mere inflammation of the chorio-decidual interface is mentioned as a primum movens producing a cascade of cytokines that result in an inflammatory response [41].

Therefore the question is: when gestational inflammation is not bacterial- or viral-induced, where does it come from?

4. Chromosomal abnormalities and genetic inflammatory polymorphisms as a cause of pregnancy inflammation and coagulation

In the absence of chronic maternal inflammatory dysease, the cases with IL-6 rise preceding infection would speak in favor of a functional inflammation arising from the fetus itself. In such cases, based on the above mentioned role of extravillous trophoblast in the control of the local maternal cellular immune response, it can be admitted that the shift from TH-1 to TH-2 type of maternal immunity, normally resulting from the fetal release of an adequate amount and quality of TH-2 mediators, did not take place, due to a failure of the trophoblast to correctly balance cytokines.

In order to confirm this opinion, it was right to investigate the physiological modulators of vascular functions and coagulation, as well as the behavior of cytochines and prostanoids involved in inflammation and smooth muscle contraction, in the fetal compartment of pregnancies with fetal aneuploidy. The reason to chose aneuploid pregnancies was that very often they end in abortion and, therefore, an imbalance of these mediators can be expected.

Accordingly, a significantly increased level of amniotic fluid IL-6, and a decreased IL-8 level in the presence of fetal aneuploidy at the 17th week of pregnancy was registered, while IL-6 concentration was reduced in the maternal blood [42].

Morover, the comparison between euploid and aneuploid pregnancies with respect to maternal serum and amniotic fluid levels of the components of the plasminogen system, showed significantly higher serum levels of urokinase plasminogen activator and its complexed form with type-1 inhibitor in aneuploidy. In addition, in amniotic fluid, tissue plasminogen activator was significantly lower in aneuploidy, whereas type-1 inhibitor was significantly higher in the cases with minor chromosomal abnormalities. In addition, the complexed form of urokinase plasminogen activator with its type-1 inhibitor was 7,53 times higher in aneuploidy [43].

With the aim to shed light on the regulation of the vascular function in pregnancies complicated by fetal chromosomal abnormalities, the potent vasoconstrictor peptide endothelin and the proangiogenic nucleoside adenosine were investigated. Amniotic fluid levels of endothelin-1 were significantly increased in pregnancies with fetal aneuploidy at the 17th gestational week [44]. As regards the adenosine transduction cascade, it is disturbed in Trisomy 21. Indeed, compared to euploid, reduced adenosine receptors, A [1] and A(2B) expression was revealed in chorionic villi and mesenchimal cells [45]. It was suggested that these vascular anomalies may lead to fetal growth restriction, malformation and abortion, well known features of aneuploid pregnancies.

These results, indicative of an umbalance of cytokines, along with abnormalities of vascular function and coagulation in fetal aneuploidies suggest that the related gestational complications may arise from the fetus itself. They support the opinion that the high incidence of miscarriage observed in chromosomal abnormalities can be interpreted as a consequence of inflammation, vascular function impairment and coagulation.

However they may also occur in euploid pregnancy, as possible expressions of fetal genetic inflammatory polymorphisms. These, indeed, are reported to be responsible for harmful inflammatory response in those who possess them. Accordingly, it has been demonstrated that maternal polymorphisms in genes IL-10, MBL, TNFRSF6 and TGFB1 may influence susceptibility to chorioamnionitis [46].

Furthermore, polymorphisms that increase the magnitude or the duration of the inflammatory response are associated with an increased risk of preterm birth, while those decreasing the inflammatory response are associated with a lower risk [47]. Moreover, an investigation on six cytokine genes associated with inflammation, namely IL-1α, IL-1ß, IL-2, IL-6, TNF, and lymphotoxin α, led to the conclusion that common genetic variants in proinflammatory cytokine genes do increase the risk for spontaneous preterm birth [48].

5. Therapy of cytokine umbalance

5.1 Antibiotics

Once it was established that the major obstetric complications arise from an inflammatory process of maternal or fetal origin, the next step was to establish the best way to prevent and cure it.

Several clinical studies showed that ajunctive antibiotic therapy aimed at the delay of childbirth, even in absence of infection, in cases of so called 'idiopathic' threatened preterm delivery, was able to significantly prolong pregnancy [49, 50]. As some antibiotics influence the intracellular level of calcium [51] and phospholipase A2 [52], a calcium-dependent enzyme involved in the regulation of prostaglandin biosynthesis, the question of their possible direct anti-inflammatory action (i.e. independent of the antibacterial effect) was raised. Therefore the effect of ampicillin on the amnionic prostaglandin E2 release was tested, showing a significant dose dependent inhibitory action of the drug on the prostanoid output. Such a result suggested that its use in the therapy of premature labor is authorized even in the absence of infection [53]. Subsequently the inhibitory action of beta-lactamines was compared with that of other classes of antibiotics. Interestingly, it was found that Ceftriaxone and, to a lesser extent, Gentamicin significantly and reversibly inhibit both basal and arachidonic acid- or oxytocin-stimulated amnionic prostaglandin E release. On the contrary, Tetracycline and Erythromycin do not influence prostaglandin E output. The inhibitory effect of ampicillin is potentiated, in an additive manner, by Ceftriaxone, reduced by Gentamycin, and eliminated by Tetracycline and Erythromycin [54]. A further relevant aspect emerges from the research on the novel action of antibiotics mentioned above: the influence of ampicillin on IL-6, one of the TH-1 cytokines able to stimulate Prostaglandin E2 release. The effect of the drug was tested on amnion-like Wistar Institute Susan Hayflick (WISH) cells as well as in amniotic fluid of patients submitted to genetic amniocentesis during the 17th week of their singleton physiological pregnancy. At doses ranging from 10^{-7} to 10^{-4} M, ampicillin decreased IL-6 as well as PGE2 release from WISH cells. Moreover, IL-6 amniotic fluid levels sampled 4 hours after ampicillin administration proved significantly and strongly reduced when compared with those sampled either before or 12 hours after treatment [55].

The effects of the above antibiotics shed new light on their utility not only in the therapy of infection, but also in inflammatory conditions, which often precede it. Indeed, contrary to the widespread delay in the use of antibiotics for fear that they may favor the appearance of resistant bacterial strains (an event limited to particular cases, which can be solved by replacing the drug), those antinflammatory effects support the indication for the early or even preventive use of beta-lactamines, in order to suppress inflammation. Therefore, to this purpose, beta-lactamines can represent the first choise, aimed at preventing the infection of the inflammed tissue. It could be argued that there are other steroidal and non-steroidal drugs to fight inflammation. It's true. But inflammation has many little-known aspects, and here one talks about the one that precedes and favors the infection, in which the anti-inflammatory action of antibiotics would logically seem preferable to that of drugs without bactericidal activity.

Based on the above considerations, the preventive use of beta-lactamines in invasive prenatal diagnosis was introduced four decedes ago at the obstetrical departement of Ferrara University Hospital. Indeed, amnicentesis, cordocentesis and chorionic villous biopsy, like any other surgical procedure, produce inflammation in the injured tissues: myometrium, decidua amnio-chorial membranes, placenta, obviously resulting in the release of TH1 cytokines and prostaglandins. The reason for choosing beta-lactamines is that they are more effective in reducing IL-6 and PGE2 release compared to other classes of antibiotics, and the in vitro and in vivo evidence already obtained was certainly sufficient to begin with. Later on, the first randomized controlled clinical trial showing the efficacy of antibiotic administration before amniocentesis in reducing the incidence of abortion and premature birth was published [56]. Azithromycin was used in this trial, probably due to its wide bactericidal effect. However it was subsequently shown that in pregnant rats the drug reduces the level of tumor necrosis factor TNF- α and increases that of IL-10, two cytokines with inflammatory and anti-inflammatory action, respectively [57]. It should be noted that the article suggests the use of azithromycin to prevent pregnancy loss 'infection- or endotoxin-dependent'. However, as it is shown in the Fetal Inflammatory Response Syndrome, inflammation may represent the condition preceding, and also leading to infection. Considering for instance the possible presence of predisposing factors to inflammation, like fetal or maternal genetic inflammatory polymorphisms, antibiotic use should not be limited to cure infection, but is also indicated to prevent and cure the preceding inflammation. It is our clinical opinion that the risk of producing resistant bacterial strains is overestimated, and therefore in half a century experience our strategy for prevention of pregnancy loss largely included preventive use of antibiotics.

5.2 Lactoferrin

Nevertheless, alternative drugs may be used in the protection of pregnancy, when an antinflammatory action is indicated without the need of a anti-bacterial one. To this purpose we tested Lactoferrin (LF), an iron-binding glycoprotein with anti-inflammatory properties, which is normally present in human organism and is largely prescribed to cure anemia. We first reported that a vaginal compound containing 300 mg of LF, administered 4 hours before genetic amniocentesis, significantly decreases amniotic IL-6 concentration [58]. Subsequently we found that the same dose of the compound significantly down-regulates 17 pro-inflammatory amniotic cytokines among which IL-9, IL-15, IFN-γ, IP-10, TNF-α, IL-1α and MCP-3, while it up-regulates several among anti-inflammatory [59]. We also evaluated the effect of vaginal LF on amniotic fluid PGE2 level and MMP-TIMP system. We found that vaginal lactoferrin significantly lowers PGE 2, active MMP-9, and its inhibitor TIMP-1. Conversely, active MMP-2 and MMP-2/TIMP-2 molar ratio are increased, whilst TIMP-2 remains unchanged [60].

5.3 Glucocorticoids

Once recognized that the majority of relevant pregnancy complications are triggered by an inflammatory process, the preventive and curative role of glucocorticoids has been better clarified. The physiologic adrenal gland circadian production of glucocorticoids represents the first defense against inflammation, throughout the corticosteroid control of the mediators of cellular functions among which IL-1,

IL-6, IL-8, Tumor necrosis factor, granulocyte-macrophage colony-stimulating factor (G-CSF), monocyte chemotactic protein-1 (MCP-1) [61]. The complex action of glucocorticoids (GCs) is exerted also on cellular cytokine receptors, which are increased in some cell types, decreased in others [62, 63]. Examples of the regulatory actions of GCs are down-regulation of the expression of the cellular receptors that recognize a variety of pathogens (Toll-like receptors) [64], as well as suppression of pro-inflammatory and up-regulation of anti-inflammatory cytokines. This effect has been reported for dexamethasone in primary isolated murine liver cells [65]. Glucocorticoid inhibition of the human pro-IL-lβ gene by decreasing DNA binding of transactivators to the signal-responsive enhancer has been shown as well [66].

An important example of the complexity of these regulatory processes to be considered is that the glucocorticoid receptor (GR) can decrease TNF stimukated IL-6 transcription independently from GCs, as a protective mechanism against excessive inflammation [67]. Moreover GCs are reported to induce, rather than to inhibit, the secretion of the migration inhibitory factor (MIF) [68], thus counteracting its own inhibition of pro-inflammatory cytokine production.

In addition to the intricate network of stimulatory and inhibitory messengers and tissue distribution of receptors, the action of the GCs is subordinated to the enzyme that transforms cortisol into cortisone, thus inactivating it: 11-beta hydroxysteroid Dehydrogenase [69]. It is widely distributed in the uterus and placenta, in immune cells, skeletal muscle and heart, while it is reported to apparently lack in the fetal organism up to the advanced stages of its development. What can this lack possibly mean? Well, the first logical implication is that normal embryonic development does not fear the effect of cortisone up to the advanced stage of its maturation.

At this regard, there is one important point to clarify to the benefit of mainstream obstetrics, and it is the difference between 'maturation' and 'inflammation'. The concept of an improvement of fetal lung maturation by betamethasone was first expressed fifty years ago, to explain the decreased incidence of 'Hyaline Membrane Disease' of neonates following the hormone administration to their mothers the day before premature birth [70].

At that time the devastating influence of inflammation on pregnancy had not been sufficiently explored, and the 'Fetal Systemic Inflammatory Response Syndrome' had not been described. Therefore it was believed that the hyaline membrane disease was caused by prematurity, and the action of betamethasone was to induce a sort of pulmonary maturation. But today, the features of fetal inflammation leading to Hyaline Membrane Disease, Necrotising Enterocholitis and Encephalopathy are well known, and therefore to keep on talking of 'maturation' instead of inflammation, it is not only an incorrect opinion: it is also misleading. Indeed, such a misinterpretation impairs the correct preventive use of GCs throughout pregnancy. A deep update is therefore needed in order to renew the guidelines on a clinical basis rather than a mere statistical one, as usally done.

Further example of the somewhat contradictory reciprocal influence of the mediators of inflammation, is that the pro-inflammatory cytokines, IL-1 and TNF-α included, up-regulate 11-β-hydroxysteroid-dehydrogenase mRNA in different cell types. Finally, GCs themselves stimulate the enzyme, appearently as an attempt to impaire their own anti-inflammatory effect.

Such complex influences are of particular relevance in understanding the nature of a balanced protective action against pregnancy loss. They indicate that the behavior of GCs in the regulation of inflammatory processes is far more complex than our limited knowledge can imagine: once recognized the number and function of the involved mediators, it is impossible to establish the precise order of their activation. The clinical protective action of glucocorticoids can only be assessed by the *'ex juvantibus'* criterion: that is, case by case, from the benefit obtained following their administration.

6. Other therapies

The COVID pandemic found health systems around the world unprepared. The technical-scientific committees of epidemiologists and virologists failed to consider therapeutic strategies, limiting the advice only to preventive measures, such as face masks, lockdown and quarantena. However, alongside hundreds of thousands of dead there have been happy islands where patients have been properly cared. Hyperimmune serum transfusions from recovered patients proved effective in saving many human lives [71]. Their efficacy depends on a direct neutralization of the virus, by preventing its entry into the cell. Attempts have been made to reduce the level of IL-6 by administering its antagonist 'tocilizumab' [72]. However, to look for a single drug capable of balancing the intricate network of stormy cytokines is a legitimate but naive hope: lowering the level of just one cytokine while that of many others remains high does not make much sense. Therefore the attention of researchers should turn to drugs capable of restoring the balance of cytokines as a whole, reducing the level of the inflammatory ones and increasing that of the anti-inflammatory, as happens with cortisone and lactoferrin.

A similar approach recently suggested the use of α-1-antitrypsin (AAT), a serine protease inhibitor providing a defense against the digestion of healty tissue by proteolitic enzymes. Interestingly, AAT blood level is very high during inflammation, as well as in advanced pregnancy, while its deficiency causes inflammation and viral infections. AAT therapy has been appruved for treatment of chronic obstructive pulmonary disease [73], and there is no reason not to test it, even as a preventive measure, in a serious emergency as that of the current pandemic.

7. Pregnancy and COVID 19 *'cytokine storm'*

During the first few years of my residency, cases of unrecognized 'pregnancy cytokine storm' were not uncommon. The pathological condition in which they occurred, in advanced gestational age, was called *'gestosis syndrome'*. Today it is improperly called *pre-eclampsia*, due to a possible complication (rare, and not the worst): tonic–clonic convulsions. More common are the sequelae of vascular pathology: *abruptio placentae*, and disseminated intravascular coagulation. These are the consequence of the inflammation triggered by the cytokine unbalance, that, once become extreme, is called 'storm'. In more advanced Obstetric Units, these ominous complications virtually disappeared because their premises are identified and taken care of before the onset of cytokine stormy release. This represents the rationale of low dose betamethasone therapy throughout the entire course of pregnancy for preventing pregnancy loss and elated complications [74–77]. Conversely, when cytokines trigger intravasal coagulation at the utero-placental level, the fetus dies, just as an adult dies from pulmonary vessels coagulation triggered by COVID 19 cytokine storm. Indeed, both deaths are caused by suffocation, because the placenta is the lung through which the fetus breathes.

A virus does not kill by itself: it does so through inflammation and coagulation, two perfectly curable pathologies as long as they are treated in time, that is, at their first onset. Unfortunately, in the management of COVID-19 pandemic, Health Services, overlooking the pathogenesis, focused on preventive measures rather than cure the disease.

Filtering facepiece respirators (FFRs) as well common face masks were core of the world health strategy. However, there is little evidence that by wearing a medical mask and washing hands provides significant protection against COVID 19 contagion. To the best of our knowledge, there is no randomized controlled clinical trial that demonstrate the efficacy of face masks in preventing the contagion. After all, it is logical to observe that the masks are not watertight, and therefore viruses can escape around everywhere. The belief of a possible efficacy, derived from the 123 years old '*Flugge's droplets*' account [78], still ignores that the virus remaines viable in aerosols over 3 hours [79], and therefore a delayed infection is likely to occur even long after a loose interaction with a carrier.

In addition, there are many unresolved questions regarding the spread of this disease. The theory of the '*patient number 1*', in Italy at first identified at Codogno (Lombardia), was nullified by the demonstration of the presence of anti-COVID-19 antibodies in the blood of healthy donors collected before the start of the pandemic. This observation suggests not only that the spread of the virus occurs in a silent way, but also that the virus is not able by itself to produce a deadly disease. For that to happen, concurrent pathologic conditions are required, some of which are well known, some others still unknown.

As reported above, the free transmission of viruses through the air, as well through other routes is well known, and scientifically confirmed. Obviously, the mere presence of viruses does not necessarily imply that the carriers must get sick: it is the well known condition of 'healty carriers'. On the other hand their absence does not exclude that the viruses can meet the same subjects in later periods of their life. The onset of the disease requires the concurrence of a compromised immune response.

In addition to the above considerations on the free and unrele nting circulation of all viral particles, COVID-19 included, the Italian experience in the unsuccessful management of the pandemic also speaks against the effectiveness of medical masks. In Italy, from North to South and all the way to the islands, everyone was forced to wear a mask, but the large majority of deaths were concentrated in four of the northern Regions: Piemonte, Lombardia, Veneto and Emilia Romagna. These are the most industrialized, rich and polluted Italian regions, which are regarded soundest as far the Italian healthcare system is concerned. However, precisely those regions reported the highest death rates, compared to all other Italian regions and other nations as well. In spite of wearing medical masks, a large number of Italian physicians and health workers died, most of which in the above mentioned regions. Further to that, at the beginning of the lockdown, a few hundred Italian citizens fled from north to south of the country, being accused of spreading the infection in the southern regions, but this did not happen at all. A surveillance study performed among healthcare workers at the 'Infectious Diseases Cotugno Hospital' in Naples, showed a very low prevalence of the COVID-19 infection among health care workers: the reason was that healthy subjects scrupulously follow protected and obligatory paths, and wear overalls and helmets that completely isolate them from the surrounding environment full of viruses [80].

The efficacy of the measures adopted at Cotugno Hospital therefore explains the little or no utility of simple masks in preventing the contagion. At the same time, Cotugno's experience demonstrates that concentrating a large number of carriers of high viral load in a limited space, without adopting the correct precautions is a serious mistake. How to proliferate subjects with a high viral load? Simple: instead of treating adequately the early symptoms of illness, leave them at home a few weeks with high fever and without effective medical treatment, then hospitalize them when the high viral load boosts inflammation up to the level of asphyxia due to pulmonary thrombosis.

8. Concluding remarks

Death from COVID 19 infection reiterates the same pathogenic mechanism of fetal and maternal death in pregnancy: it is a matter of inflammation triggered by umbalanced cytokines and coagulation in the lung, the same that happens in pregnancy, starting at the utero-placental level. In the first the cause is the virus, in the second is the fetus itself, as it is explained in the above reported literature.

In both cases, the cause cannot be eliminated.

The cure, instead, exists: it is the same in both conditions and is very effective. The rationale for management is not to fight the cause, but to cure the disease, i.e. inflammation, and consequent overlapping bacterial infection and thrombosis.

Therapeutic agents include cortison and eventually other non-steroidal drugs against inflammation, antibiotics against superimposed bacterial infection and heparin against thrombosis. However, it can be stated: no cytokine umbalance = no inflammation, no inflammation = no infection = no intravascular coagulation. Moreover, it must be stressed that the treatment is all the more effective the earlier it is started. The same therapy that can be effective if started at the first onset of symptoms, becomes 'compassionate' if started when inflammation and thrombosis are already in an advanced stage [81, 82].

Author details

Fortunato Vesce
Obstetric and Ginecology Unit, Ferrara University, Italy

*Address all correspondence to: ves@unife.it

References

[1] Biondi C, Pavan B, Dalpiaz A, Valerio A, Spisani S, Vesce F (2005) Evidence for the presence of N-formyl-methionyl-leucyl-phenylalanine (fMLP) receptor ligands in human amniotic fluid and fMLP receptor modulation by physiological labour. J Reprod Immunol 68:71-83.

[2] Biondi C, Pavan B, Ferretti ME, Corradini FG, Neri LM, Vesce F (2001) Formyl-methionyl-leucyl-phenylalanine induces prostaglandin E2 release from human amnion-derived WISH cells by phospholipase C-mediated [Ca+]i rise. Biol Reprod 64:865-70.

[3] Buzzi M, Vesce F, Ferretti ME, Fabbri E, Biondi C (1999) Does formyl-methionyl-leucyl-phenylalanine exert a physiological role in labor in women? Biol Reprod 60:1211-6.

[4] Lunghi L, Ferretti ME, Medici S, Biondi C, Vesce F (2007) Control of human trophoblast function. Reprod Biol Endocrinol Feb 8;5:6.

[5] Raghupathy R (1997) Th1-type immunity is incompatible with successful pregnancy. Immunol Today 18:478-82.

[6] Raghupathy R, Al-Azemi M, Azizieh F (2012) Intrauterine Growth Restriction: Cytokine Profiles of Trophoblast Antigen-Stimulated Maternal Lymphocytes. Clin Dev Immunol 734865.

[7] Al-Azemi M, Raghupathy R, Azizieh F (2017) Pro-inflammatory and anti-inflammatory cytokine profiles in fetal growth restriction. Clin Exp Obstet Gynecol 44:98-103.

[8] Medawar PB (1953) Some immunological and endocrinological problems raised by the evolution of viviparity in vertebrates. Symp Soc Exp Biol 44:320-38.

[9] Billington WD (2003) The immunological problem of pregnancy: 50 years with the hope of progress. A tribute to Peter Medawar. J Reprod Immunol 60:1-11.

[10] Luppi P, Haluszczak C, Betters D, Richard CA, Trucco M, DeLoia JA (2002) Monocytes are progressively activated in the circulation of pregnant women. J Leukoc Biol 72:874-84.

[11] Kraus TA, Sperling RS, Engel SM, Lo Y, Kellerman L, Singh T, et al. (2010) Peripheral blood cytokine profiling during pregnancy and post-partum periods. Am J Reprod Immunol 64:411-26.

[12] Aghaeepour N, Ganio EA, McIlwain D, Tsai AS, Tingle M, Van Gassen S, et al. (2017) An immune clock of human pregnancy. Sci Immunol 2:aan2946.10.1126/sciimmunol.aan2946.

[13] Gomez-Lopez N, Romero R, Hassan SS, Bhatti G, Berry SM, Kusanovic JP, et al. (2019). The cellular transcriptome in the maternal circulation during normal pregnancy: a longitudinal study. Front Immunol 10:2863. 10.3389/fimmu.2019.02863.

[14] Forger F, Villiger PM (2020) Immunological adaptations in pregnancy that modulate rheumatoid arthritis disease activity. Nat Rev Rheumatol 16:113-22. 10.1038/s41584-019-0351-2.

[15] Gomez-Lopez N, Romero R, Xu Y, Miller D, Leng Y, Panaitescu B, et al. (2018) The immunophenotype of amniotic fluid leukocytes in normal and complicated pregnancies. Am J Reprod Immunol 79:e12827. 10.1111/aji.12827.

[16] Sacks GP, Studena K, Sargent K, Redman CW (1998) Normal pregnancy and preeclampsia both produce inflammatory changes in peripheral

blood leukocytes akin to those of sepsis. Am J Obstet Gynecol 179:80-6.

[17] Redman CW, Sacks GP, Sargent IL (1999) Preeclampsia: an excessive maternal inflammatory response to pregnancy. Am J Obstet Gynecol 180:499-506.

[18] Naccasha N, Gervasi MT, Chaiworapongsa T, Berman S, Yoon BH, Maymon E, et al. (2001) Phenotypic and metabolic characteristics of monocytes and granulocytes in normal pregnancy and maternal infection. Am J Obstet Gynecol 185:1118-23.

[19] Sugiura-Ogasawara M, Nozawa K, Nakanishi T, Hattori Y, Ozaki Y (2006) Complement as a predictor of further miscarriage in couples with recurrent miscarriages. Hum Reprod 21:2711-4.

[20] Hiby SE, Apps R, Sharkey AM, Farrell LE, Gardner L, Mulder A, et al. (2010) Maternal activating KIRs protect against human reproductive failure mediated by fetal HLA-C2. J Clin Invest 120:4102-10.

[21] Yao Y, Xu XH, Jin L (2019) Macrophage polarization in physiological and pathological pregnancy. Front Immunol 10:792.

[22] Yang F, Zheng Q, Jin L (2019) Dynamic function and composition changes of immune cells during normal and pathological pregnancy at the maternal-fetal interface. Front Immunol 10:2317.

[23] Pierik E, Prins JR, van Goor H, Dekker GA, Daha MR, Seelen MAJ, et al (2019) Dysregulation of complement activation and placental dysfunction: a potential target to treat preeclampsia? Front Immunol 10:3098.

[24] Mertz D, Geraci J, Winkup J, Gessner BD, Ortiz JR, Loeb M (2017) Pregnancy as a risk factor for severe outcomes from influenza virus infection: a systematic review and meta-analysis of observational studies. Vaccine 35:521-8.

[25] Cornish EF, Filipovic I, Asenius F, Williams DJ, McDonnel T (2020) Innate Immune Responses to Acute Viral Infection During Pregnancy. Front Immunol 11: 572567.

[26] Giakoumelou S, Wheelhouse N, Cuschieri K, Entrican G, Howie SEM, Horne AW (2016) The role of infection in miscarriage. Hum Reprod Update 22:116-133.

[27] Donders GG, Van Bulck B, Caudron J, Londers L, Vereecken A, Spitz B (2000) Relationship of bacterial vaginosis and mycoplasmas to the risk of spontaneous abortion. Am J Obstet Gynecol 183:431-437.

[28] Srinivas SK, Ma Y, Sammel MD, Chou D, McGrath C, Parry S, Elovitz MA (2006) Placental inflammation and viral infection are implicated in second trimester pregnancy loss. Am J Obstet Gynecol 195:797-802.

[29] Mazzoni E, Pellegrini E. Mazziotta C, Lanzillotti C, Rotondo JC, Bononi I, Iaquinta MR, Manfrini M, Vesce F, Tognon M, Martini F (2020) Mother to child transmission of oncogenic polyomaviruses BKPyV, JCPyV and SV40. J inf 2020.02006.

[30] Tognon M, Tagliapietra A, Magagnoli F, Mazziotta C, Oton-Gonzalez L, Carmen Lanzillotti C, Vesce F, Contini C, Rotondo JC, Martini F (2020) Investigation on Spontaneous Abortion and Human Papillomavirus Infection. Vaccines (Basel) 2020 Sep; 8(3): 473. Published online 2020 Aug 25. doi: 10.3390/vaccines8030473.

[31] Di Giulio DB (2012) Diversity of microbes in amniotic fluid. Semin Fetal Neonatal Med 17:2-11.

[32] Baschat AA, Towbin J, Bowles NE, Hamman CR, Weiner CP (2003)

Precalence of viral DNA in amniotic fluid of low-risk pregnancies in the second trimester. J Matern Fetal Neonatal Med 13:381-4.

[33] Wenstrom KD, Andrews WW, Bowies NE, Towbin JA, Hauth JC, Goldberg RL (1998) Intrauterine viral infection at the time of second trimester genetic amniocentesis. Obstet Gynecol 92:420-4

[34] Miller JL, Hamman C, Weiner C, Baschat AA (2009) Perinatal outcomes after second trimester detection of amniotic fluid viral genome in asymptomatic patients. J Perinat Med 37:140-3.

[35] Cardenas I, Means RE, Aldo P, Koga K, Lang SM, Booth CJ, et al. (2010) Viral infection of the placenta leads to fetal inflammation and sensitization to bacterial products predisposing yo preterm labor. J Immunol 185:1248-57.

[36] Romero R, Gomez R, Ghezzi F, Yoon BH, Mazor M, Edwin SS, Berry SM (1998) A fetal systemic inflammatory response is followed by the spontaneous onset of preterm parturition. Am J Obstet Gynecol 179:186-93.

[37] Gomez R, Romero R, Ghezzi F, Yoon BH, Mazor M, Berry SM (1998) The fetal inflammatory response syndrome. Am J Obstet Gynecol 179:194-202.

[38] Adams Waldorf KM, Rubens CE, Gravett MG (2011) Use of nonhuman primate models to investigate mechanisms of infection-associated preterm birth. BJOG 118, 136-144.

[39] Mendz GL, Kaakoush NO, Quinlivan JA(2013) Bacterial aetiological agents of intra-amniotic infections and preterm birth in pregnant women. Front Cell Infect Microbiol 3: 58.

[40] Keelan JA, Blumenstein M, Helliwell RJ, Sato TA, Marvin KW, Mitchell MD (2003) Cytokines, prostaglandins and parturition--a review. Placenta 24:33-46.

[41] Muglia LJ, Katz M (2010) The enigma of spontaneous preterm birth. N Engl J Med 362:529-35.

[42] Vesce F, Scapoli C, Giovannini G, Tralli L, Gotti G, Valerio A, Piffanelli A (2002) Cytokine imbalance in pregnancies with fetal chromosomal abnormalities. Hum Reprod 17:803-8.

[43] Vesce F, Scapoli C, Giovannini G, Piffanelli A, Geurts-Moespot A, Sweep FC (2001) Plasminogen activator system in serum and amniotic fluid of euploid and aneuploid pregnancies. Obstet Gynecol 97:404-8.

[44] Vesce F, Farina A, Jorizzo G, Tarabbia C, Calabrese O, Pelizzola D, Giovannini G, Piffanelli A (1996) Raised level of amniotic endothelin in pregnancies with fetal aneuploidy. Fetal Diagn Ther 11:94-8.

[45] Gessi S, Merighi S, Stefanelli A, Mirandola P, Bonfatti A, Fini S, Sensi A, Marci R, Varani K, Borea PA, Vesce F (2012) Downregulation of A(1) and A(2B) adenosine receptors in human trisomy 21 mesenchymal cells from first-trimester chorionic villi. Biochim Biophys Acta 1822:1660-70.

[46] Annells MF, Hart PH, Mullighan CG, Heatley SL, Robinson JS, McDonald HM (2005) Polymorphisms in immunoregulatory genes and the risk of histologic chorioamnionitis in Caucasoid women: a case control study BMC Pregnancy and Childbirth volume 5, Article number: 4.

[47] Holst D , Garnier Y (2008) Preterm birth and inflammation-The role of genetic polymorphisms. Eur J Obstet Gynecol Reprod Biol 141:3-9.

[48] Engel SAM , Erichsen HC, Savitz DA, Thorp J, Chanock SJ, Olshan

AF (2005) Risk of spontaneous preterm birth is associated with common proinflammatory cytokine polymorphisms. Epidemiology 16:469-77.

[49] Mac Gregor JA, French JI, Reller LB, Todd JK, Malowski EI (1986) Adjunctive erythromicin treatment for idiopathic preterm labor: results of a randomized, double-blinded, placebo-controlled trial. Am J Obstet Gynecol 154:98-103.

[50] Morales WJ, Angel JL, O Brien WF, Knuppel LA, Finazzo M (1988) A randomized study of antibiotic therapy in idiopathic preterm labor. Obstet Gynecol 72:829-33.

[51] Burroughs SF, Johnson GJ (1993) Beta-lactam antibiotics inhibit agonist-stimulated platelet calcium influh. Thromb Haemostat 69:503-8.

[52] Tanaka K, Itazaki H, Yoshida T (1992) Cinatrins, a novel family of phospholipase A2 inhibitors. II. Biological activities. J Antibiot Tokyo 15:50-6.

[53] Vesce F, Buzzi M, Ferretti ME, Pavan B, Bianciotto A, Jorizzo G, Biondi C (1998) Inhibition of amniotic prostaglandin E release by ampicillin. Am J Obstet Gynecol 178:759-64.

[54] Vesce F, Pavan B, Buzzi M, Pareschi MC, Bianciotto A, Iorizzo G, Biondi C (1999) Effect of different classes of antibiotics on amniotic prostaglandin E release. Prostaglandins Other Lipid Mediat 57:207-18.

[55] Vesce F, Pavan B, Lunghi L, Giovannini G, Scapoli C, Piffanelli A, Biondi C (2004) Inhibition of amniotic interleukin-6 and prostaglandin E2 release by ampicillin. Obstet Gynecol 103:108-13.

[56] Giorlandino C, Cignini P, Cini M, Brizzi C, Caccioppolo O, Milite V, Coco C, Gentili P, Mangiafico L, Mesoraca A, Bizzoco D, Gabrielli I, Mobili L (2009) Antibiotic profilaxis before second-trimester generic amniocentesis(APGA): a single-centre open randomised trial. Prenat Diagn 29:606-12.

[57] Ayse Er (2013) Azithromycin prevents pregnancyloss: reducing the level of tumor necrosis factor-alfa and raising the level of interleukin-10 in rats. Mediators Inflamm 2013:928137.

[58] Vesce F, Giugliano E, Bignardi S, Cagnazzo E, Colamussi C, Marci R, Valente N, Seraceni S, Maritati M, Contini C (2014) Vaginal lactoferrin administration before genetic amniocentesis decreases amniotic interleukin-6 levels. Gynecol Obstet Invest 77:245-9.

[59] Maritati M, Comar M, Zanotta N, Seraceni S, Trentini A, Corazza F, Vesce F, Contini C (2017) Influence of vaginal lactoferrin administration on amniotic fluid cytokines and its role against inflammatory complications of pregnancy. J Inflamm (Lond) Feb 15;14:5. doi: 10.1186/s12950-017-0152-9. eCollection 2017.

[60] Trentini A, Maritati M, Cervellati C, Manfrinato MC, Gonelli A, Volta CA, Vesce F, Greco P, Dallocchio F, Bellini T, Contini C (2016) Vaginal Lactoferrin Modulates PGE_2, MMP-9, MMP-2, and TIMP-1 Amniotic Fluid Concentrations. Mediators Inflamm 2016:3648719.

[61] Fantuzzi G, Ghezzi P (1993) Glucocorticoids as cytokine inhibitors: role in neuroendocrine control and therapy of inflammatory disease. Mediators Inflamm 2: 263-270.

[62] Akahoshi T, Oppenheim JJ, Matsushima K (1988) Induction of high-affinity Interleukin-1 receptor on human peripheral blood hepathocytes by glucocorticoid hormones. J Exp Med 167:924-936.

[63] Gottschall PE, Koves K, Mizuno K, Tatsuno I, Arimura A (1991)

Glucocorticoid upregulation on interleukin-1 receptor expression in a glioblastoma cell line. Am J Physiol 261: E362-E368.

[64] Chinenov Y, Rogatsky I (2007) Glucocorticoids and the innate immune system: crosstalk with the toll-loke receptor signalling network. Mol Cell Endocrinol 275:30-42.

[65] Broering R, Montag M, Jiang M, Lu M, Sowa JP, Kleinehr K, Gerken G, Schlaak JF (2011) Corticosteroids shift the Toll-Like receptor response pattern of primary isolated murine liver cells fro an inflammatory to an anti-inflammatory state. Int Immunol 23:537-44.

[66] Waterman WR, Xu LL, Tetradis S, Motyckova G, Tsukada J, Saito K, Webb AC, Robinson DR, Auron PE (2006) Glucocorticoid inhibits the human pro-interleukin 1 beta gene (IL1B) by decreasing DNA binding of transactivators to the signal-responsive enhancer. Mol Immunol 43:773-82.

[67] Verhoog NJ, Du Toit A, Avennant C, Hapgood JP (2011) Glucocorticoid-independent repression of tumor necrosis factor (TNF) alpha-stimulated interleukin (IL)-6-expression by the glucocorticoid receptor: a potential mechanism for protection against an excessive inflammatory response. J Biol Chem 286:19297-310.

[68] Calandra T, Bernhagen J, Metz CN, Spiegel LA, Bacher M, Donnelly T, Cerami A, Bucala R (1995) MIF as a glucocorticoid-induced modulator of cytokine production. Nature 377:68-71.

[69] Chapman K, Holmes M, Seckl J (2013) 11-beta hydroxysteroid Dehydrogenases: intracellular Gate-Keepers of tissue glucocorticoid action. Physiol Rev 93(3):1139-1206.

[70] Liggins GC, Howie RN (1972) A controlled trial of antepartum glucocorticoid treatment for prevention of the respiratory distress syndrome in premature infants. Pediatrics 50:515-25.

[71] Guo, G, Ye L, Pan K, Chen Y, Xing D, Yan K, et al. (2020) New insight of emerging SARS-COV-2: epidemiology, etiology, clinical features, clinical treatment and prevention. Front Cell Dev Biol 8;410.

[72] Guaraldi G, Meschiari M, Cozzi-Lepri A, Milic J, Tonelli R, Menozzi M, et al. (2020) Tocilizumab in patients with severe COVID-19: a retrospective cohort study. Lancet Rheumatol 2, e474-84.

[73] Martini F, De Mattei M, Contini C, Tognon M (2020) Potential use of Alpha-1 Anti-trypsin in the COVID-19 treatment. Frontiers in Cell and Develop Biol 8: Art 577528 doi:10.3389/cell.2020.577538.

[74] Lunghi L, Pavan B, Biondi C, Paolillo R, Valerio A, Vesce F, Patella A (2010). Use of glucocorticoids in pregnancy. Curr Pharm Des 16:3616-37.

[75] Vesce F, Giugliano E, Cagnazzo E, Bignardi S, Mossuto E, Servello T, Marci R (2012) The role of glucocorticoids in pregnancy: four decades experience with use of betamethasone in the prevention of pregnancy loss. In "Glucocorticoids. New recognition of our familiar friend". Edited by Xiaoxiao Qian, INTECH, November 2012, Croatia, pag 407-48.

[76] Vesce F, Cagnazzo E, Giugliano E, Mossuto E, Marci R (2014) The behaviour of the peripheral natural killer cells in the foetal growth restriction. Eur Rev Med Pharmacol Sci 18:2248-52.

[77] Vesce F, Giugliano E, Cagnazzo E, Mossuto E, Marci R (2014) Low dose of betamethasone throughout the whole course of pregnancy and fetal growth: a clinical study. Eur Rev Med Pharmacol Sci 18(4):593-8.

[78] Flügge C (1897) Ueber die nachsten Aufgaben zur Erfor- schung der Verbreiturigs weise der Phtisie, Deut Med Wochen, 14 oct. 1897 p. 665.

[79] van Doremalen N, Bushmaker T, Morris DH, et al (2020) Aerosol and Surface Stability of SARS-CoV-2 as Compared with SARS-CoV-1. N Engl J Med 382:1564-1567.

[80] Fusco FM, Pisaturo M, Iodice V, et al. (2020) COVID-19 among healthcare workers in a specialist infectious diseases setting in Naples, Southern Italy: results of a cross-sectional surveillance study. J Hosp Infect 105:596-600.

[81] Contini C, Caselli E, Martini F, Maritati M, Torreggiani E, Seraceni S, Vesce F, Perri P, Rizzo L, Tognon M (2020) COVID 19 is a multifaceted challenging pandemic which needs urgent public health interventions. Microorganisms 8, 1228.

[82] Derwand R, Scholz M, Zelenko V (2020) COVID-19 outpatients: early risk-stratified treatment with zinc plus low-dose hydroxychloroquine and azithromycin: a retrospective case series study. International Journal of Antimicrobial Agents 56; 106214.

Section 3

Cancer and Injuries

Chapter 6

IL-17 Biological Effects and Signaling Mechanisms in Human Leukemia U937 Cells

Samuel Evans Adunyah, Richard Akomeah, Fareed K.N. Arthur, Roland S. Cooper and Joshua C.M. Williams

Abstract

Human Interlekin-17 is produced by memory activated CD4+ T cells and other cells. It was initially considered unique in that its specific receptor is distinct from other cytokine receptors. IL-17 receptor is ubiquitously expressed by different cells including T cells. IL-17 plays a role in regulating growth, immune response and pro-inflammatory responses. It regulates differentiation of a subset of Th0 cells into Th-17 cells, which produce IL-17-induced cytokines. The IL-17R belongs to type 1 cytokine receptors. IL-17 belongs to a superfamily of its own, which includes IL-17A, IL-17B, IL-17C, IL-17E and IL-17F. These members of IL-17 superfamily have some sequence homology but bind to different receptors. Prior to this investigation, limited information existed on the effects of IL-17A in human leukemia cell lines. Our results show that IL-17A promotes growth, anti-apoptotic effects, chemotaxis, cytokine expression and transcriptional factor activation in leukemia cells. IL-17A activates multiple signaling pathways including PI-3 K, Jak–STAT, Raf-ERK1/2 and SRC kinase pathways, which mediate different biological effects of IL-17A in leukemia cells. Our findings implicate IL-17A in leukemia cell growth and survival, supporting potential leukemia therapy via development of anti-IL-17A drugs. This chapter focuses on IL-17A, herein referred to as IL-17.

Keywords: IL-17A, leukemia, cytokines, Jak/STAT, PI-3 K/Akt, ERK1/2 transcriptional factors

1. Introduction

IL-17 is a unique cytokine which was initially discovered through differential and subtractive screening of clones from DNA library from murine lymphoid cells and initially called T-lymphocyte associated antigen 8 (CTLA-8) [1]. IL-17 was found to have 50% sequence homology to the open reading frame 13 (ORF-13) in herpes virus Saimiri [1]. Subsequently, the human homolog of CTLA-8 was identified [1] and its incubation with human fibroblast resulted in induction of both IL-6 and IL-8. This led to renaming of the CTLA-8 human equivalent as IL-17 [2]. Human IL-17 production was also found to be limited to particular cellular elements of the immune system and that activated CD4 + T cells of the Th1/Th0 subset and stimulated memory T cells, synthesize IL-17 [3, 4]. IL-17 is a glycosylated homodimeric protein of 30 to 35 kDa also produced by nickel-specific T lymphocytes and it regulates I-CAM-1 expression and chemokine production [5]. Unlike other cytokines, IL-17 was noted to bind to a unique receptor distinct from other cytokine receptors [1–3]. IL-17 is ubiquitously expressed in thymocytes activated by CD3 mAb, CD45R0+ population of T cells, CD8+ splenic cells in mouse cells and synovial fluid of patients with rheumatoid arthritis [6–8]. IL-17 is abundantly produced in CD4+ T cells, now known as Th-17 cells [9–11]. It is also expressed in human peripheral blood mononuclear cells (PBMC) in response to ocular lysate in patients with birdshot chroioretinopathy [12]. IL-17 has biological effects in many cells and tissues [13–15]. IL-17 induces expression and secretion of IL-1-beta, IL-6, IL-8. TNF, GM-CSF, G-CSF, ICAM-1, and PGE2 [16–20]. The molecular characterization of IL-17 receptor (IL-17R) was reported in 1997 [21]. IL-17R is a type 1 transmembrane receptor and it is a single chain, which shares some properties with IL-2R-beta chain, and GM-CSFR, all of which are type 1 membrane receptors [21, 22]. IL-17R is also expressed in synovial endothelial cells and Chondrocytes from arthritis patients [23, 24].

Five different IL-17 ligands are now characterized as members of IL-17 superfamily of cytokines and differ from other cytokines but share some sequence homology with each other [25]. Among the IL-17 super family, IL-17A is most commonly expressed in many tissues as well as in cells of hematopoietic origin including monocytes and macrophages. In addition to IL-17A, there are IL-17B, IL-17C, IL-17E and IL-17F [25, 26]. IL-17B, IL-17C and IL-17E expression are widespread in many tissues including testis, brain and kidney [25, 26]. Each IL-17 family members have their individual specific receptor as these IL-17 family members do not bind to the same receptor type [26–28]. The different members of IL-17 superfamily have different expression patterns but with similar abilities to stimulate cytokine effectors illustrating the potential for the members of the IL-17 superfamily to differentially regulate cellular responses in a wide variety of cells [28–31].

IL-17 regulates hematopoietic cell proliferation, immune response, pro-inflammatory responses [32–35] and activate specific types of T cells now known as Th-17 cells [9, 10]. These Th-17 cells play a role in host defense against extracellular pathogens by mediating recruitment of neutrophils and macrophages to infected tissues [9–11]. Th-17 cells secrete IL-17 cytokines, which in turn induce expression of IL-17-depedemt cytokines. Hence, aberrant regulation of Th-17 cells may play a role in the pathogenesis of multiple inflammatory and autoimmune disorders [9–11]. IL-17 promotes chemotaxis in human monocytes and regulates angiogenesis and cytokine production in endothelial cells [12, 36–38]. Although the target cells of IL-17-mediated signaling include immune cells such as neutrophils and macrophages [32–38], majority of IL-17's biological effects were seen in cells of either epithelial or mesenchymal origin [39–43]. The role of IL-17 in immunological function was initially examined in vivo in mice by overexpressing IL-17 in the liver of mice where an enhanced granulopoiesis and leukopoiesis led to an 80% increase in splenic mass [38, 39]. IL-17-induced accumulation of neutrophils in the airways requires involvement of GM-CSF [44]. Also, regulation of endogenous stem cell by IL-17 requires both GM-CSF and Stem Cell Factor (SCF) [45]. IL-17 and G-CSF are synergistically involved in the maintenance of normal granulopoiesis [45]. The IL-17R is ubiquitously expressed [12–15] and may explain the ability of IL-17 to stimulate peripheral blood stem cells in in mice [44, 45]. Also, There IL-17 plays active in vivo role in chemoattraction of cells of immune system [46, 47]. In addition, IL-17 exhibits paracrine effects in different cell types [48] whereby secreted IL-17 from T cells binds to its putative receptor on neighboring cells such as fibroblasts and trigger signaling that leads to NF-kB- mediated induction of expression and secretion of ICAM-I, IL-2, IL-6, IL-8 [16–20] and other cytokines, which produce different biological effects [45–50]. Also some of the T-cell secreted IL-17 become sequestered and neutralized by a soluble IL-17R (sIL-17R) [51].

Clinically, IL-17 is implicated in numerous diseases including arthritis [52–54], classical Hodgkin lymphomas [55–57], multiple myeloma [58–60], airway diseases including asthma [61, 62], musculoskeletal diseases [63, 64], inflammatory bowel diseases (IBDs) [65] autoimmune diseases [66], and different types of cancer [67–69]. Significantly elevated level of IL-17 and IL-17R are found in these diseases and IL-17 and IL-17R are known to promote anti-apoptotic effects and survival mechanisms in some types of cancer [55–66]. In most cases, IL-17 itself and/or IL-17-dependent cytokines produced downstream of the IL-17R, contribute to various pathological conditions associated with these diseases [55–68]. Furthermore, Hox3/IL-17R expression ratio has been implicated in poor prognosis in some breast cancer patients undergoing tamoxifen chemotherapy as IL-17 promotes resistance to chemotherapy in breast cancer [67–71]. IL-17 is also implicated in cervical and ovarian cancer [72–74]. Similarly, expression of IL-17 R-like protein has been detected in androgen independent prostate cancer cell lines and it has been implicated in conferring resistance to apoptosis and promoting prostate cancer via MM7-induced epithelial-to-mesenchymal transition [75–77]. IL-17 is implicated in CNS and other neurological diseases [78–80], and psoriasis [81–83]. Recent reports suggest potential role for IL-17 in the “cytokine storm event” seen in advance Coronavirus Disease 2019 (COVID-19) infection with inflammation, and pro-thrombic events in severe COVID-19 patients [84–87]. Hence, there is strong interests in understanding IL-17’s biology and it roles in COVID-19 patients [85–87]. It is not surprising that IL-17’s role in these diseases [60–87], have prompted experts in pharmaceutical industries to develop anti-IL-17 type therapies for diseases in which IL-17 is implicated [88–90].

Earliest report on IL-17 induced activation of MAP kinases and NF-kB pathways was made in chondrocytes [91]. Subsequently, IL-17 was shown to activate Raf-MAPK and Jak/STAT signaling pathways in leukemia cells [92, 93]. These reports show that IL-17 stimulates rapid phosphorylation of RAF, Erk-1/2, Jak1, Jak2, Jak3 and Stat1, Stat2 and Stat3 in human leukemia cells [92, 93]. Currently, IL-17 is known to activate and utilize multiple signaling pathways including the aforementioned as well as JNK, p38 and PI-3 K/Akt pathways to produce diverse biological effects [91–96]. Many reports have confirmed IL-17-induced activation of PI-3 K/Akt signaling mechanisms in both normal and transformed cells [96–99]. IL-17 signaling pathways are implicated in human diseases including inflammation and cancer [100]. Furthermore, TRAF and TGF-beta-1/smad2/3 signaling pathways are activated by IL-17 [99–102]. Most of the biological effects of IL-17 were initially observed in different variety of cells but to lesser extent in leukemia cells. Therefore, our rationale for initiating this study was to determine the biological effects of IL-17 in leukemia cells and elucidate the various signaling pathways utilized by IL-17 in leukemia cells. Furthermore, we wanted to determine which transcriptional factors are activated by IL-17. Finally, we wanted to determine whether IL-17 protects leukemia cells from undergoing apoptosis since previous report [75] indicated IL-17—mediates cancer cell resistance to apoptosis.

2. Experiment reagents and protocols

We purchased human U937 and THP-1 leukemia cell lines from American Tissue Culture Collection (ATCC) in Manassas, VA, USA. The cells were cultured in Roswell Park Memorial Institute-1640 media, which contained L-Glut (2 mM). Charcoal-filtered and frozen fetal bovine serum (FBS) was purchased from Atlanta Biological, Georgia, USA. Following careful thawing under sterile conditions we heated the serum at 55 °C for 45 minutes for inactivation. After cooling, we vacuum filtered the FBS under the hood and stored 50 ml aliquots at minus 20 °C. Both streptomycin and penicillin were purchased from Invitrogen, Carlsbad, CA. Prior to using the culture media, we added FBS to a final concentration of 10% (v/v). To prevent bacterial growth in the culture media, we added penicillin (50 U/ml) and streptomycin (50 U/ml). We cultured the cells in either 25 ml or 50 ml of complete media in tissue culture flasks in a CO_2 incubator set to 5% CO_2, 37 °C and 100% humidity. Typically, we passaged the cells 8x before starting a fresh culture. U937 leukemia cells were used in most of the experiments described here.

2.1 Monocyte isolation

We purchased de-identified human blood samples from New York Blood Center, Long Island, NY and Percoll gradients from GR Health Care, Piscataway, NJ [103]. To isolate monocytes for chemotaxis assays, we isolated peripheral blood mononuclear cells (PBMC) as previously described [103]. After centrifugation to remove all red blood cells, the white blood cells were carefully retrieved and suspended in 10 ml of complete media and spread across the surface of plastic dishes. The plastic dishes were incubated in the incubator for 1.5 hours to allow monocytes to attach to the surface of the dishes. Subsequently, all non-adherent cells were carefully aspirated off and discarded. The attached monocytes were carefully scraped from the dishes and suspended in culture media. The monocytes were about 95% pure based on positive staining for CD14 marker.

2.2 Detection of cytokine expression by cytokine antibody array

To determine the effect of IL-17 on cytokine expression in leukemia cells, we performed cytokine antibody array using tissue media from untreated and IL-17 treated leukemia cells [103]. Specifically, 20million cells were either untreated or treated with IL-17 (100 ng/ml) alone or with IL-17 (100 ng/ml) plus the PI-3 K inhibitor LY20094 (20 uM) or with LY20094 (20 uM) alone for 24 hours. The tissue culture media were filtered to remove debris and their protein concentration determined by Coomassie Blue Protein Assay Kit (Pierce, IL). The culture media containing 50 ug protein from untreated and treated cells were spotted onto each of the cytokine/antibody array membranes containing antibodies for over 42 cytokines (RayBiotech, Corners, GA, USA), and incubated with gentle shaking for 2 hours at 30 °C. This allowed hybridization of each cytokine in the media to its respective cytokine antibody on the array. Next, the media was carefully removed and each membrane was washed 5x with wash buffer (provided by kit) to remove all non-specific binding. Each membrane was incubated for a specified time in the color development solution provided with kit and air dried. The dark spots representing various cytokines were visualized and quantitated by digital image scanning. The spot intensities were converted to fold change relative to the corresponding spots on the membrane of the untreated cells, which was set as 1-fold.

2.3 Western blotting detection of proteins and phosphoproteins

One million leukemia cells per ml media were either untreated (control) or treated with [92, 93] IL-17 (1 ng/ml) for 2, 5, 15, 30, 60 min or with IL-17 (100 ng/ml) at 30 °C treatments up to 48 hours. Next, the cells were rapidly pelleted by micro-centrifugation at 1,800 x g for 3 minutes. The pelleted cells were washed 3x with PBS. The final pellets were collected by centrifugation at 1800 rpm for 3 min and each pellet was lysed in 500 ul of cell lysis buffer A (containing protease inhibitors, 0.5% Triton X-100, 50 mM NaF and 2 mM Vanadate) [7]. Total cell lysate protein concentration was determined by Coomassie Blue Protein Assay Kit (Pierce, IL) and 240 ug per sample was solubilized in 50 ul SDS gel sample buffer and resolved by 12% polyacrylamide gel electrophoresis. The protein bands were transferred to membrane and the membrane background blocked in a blocking buffer containing 5% milk. The membranes were incubated with specific antibodies to either total PI-3 K, p-PI-3 K, p-Akt^{Ser473}, p-Akt^{Thre308}, total Akt, p-STAT3, total STAT3, p-BAD, p-caspase3 or pGSK3-beta or total actin (for loading control) and protein bands detected [103]. The band intensities were scanned by digital image analyzer for quantitation and the band intensity from IL-17 treated samples compared to the intensity in the untreated sample.

2.4 Co-immunoprecipitation (co-IP) assay/Western blot

Total and phosphorylated Akt can associate with effector proteins [104, 105] via protein–protein interaction [106] and contribute to their regulation. We rationalized that if IL-17 promotes association between p-Akt and any of its downstream effectors, those proteins will be contained in the pulled down p-Akt-antibody complex and can be detected by co-IP/Western blot. To determine whether Akt/p-Akt binds to p-BAD or p-Caspase3 or p-GSK-3 (p-Akt's downstream effectors), we carried out co-IP. Specifically, 240 mg/ml protein from untreated or IL-17 treated cells were suspended in PBS (500 ul) in Eppendorf tubes and incubated with specific antibody that recognizes both total Akt and p-Akt^{Ser473} overnight with gentle shaking at 4 °C. Next, protein A agarose slurry (500 ul) was added to the complex in each tube and incubated for 2 hours at 4 °C with gentle shaking to allow protein A agarose to capture all the phosphoproteins bound to Akt/p-Akt-antibody complex. The tubes were centrifuged at 10,000 x g for 10 minutes at 4 °C, the supernatants carefully removed and the pellets were washed 5x with lysis buffer (see above). After the final wash, the phosphoprotein complexes in each tube was solubilized in SDS-gel sample buffer (50 ul) and boiled for 3 minutes to dissociate the phosphoproteins in each complex. The dissociated phosphoproteins were separated on 12% SDS-polyacrylamide gel as described above. The phosphoprotein bands were transferred to membrane and Western blotted for either p-BAD or p-Caspase3 p-GSK-3 or total Akt using specific antibodies. The band intensity for p-BAD, p-Caspase3 and p-GSK-3 were scanned with digital image analyzer. The old levels in the IL-17 treated cells calculated relative to the bands in the untreated cells. Representative results are presented.

2.5 Detection of IL-17RA in U937 leukemia cells and THP-1 leukemia cells

Specific antibody to IL-17AR was purchased from Santa Cruz, CA and goat anti-rabbit IgG-HRP antibody was from Amersham, CA. In order to detect IL-17AR, 240 ug of total cell lysate protein from 40 million cells was separated on 15% polyacrylamide gel [106]. The protein bands were transferred to nitrocellulose membrane followed by incubation in a blocking buffer containing 5% filtered non-fat milk to block non-specific sites on the membrane. Both the IL-17AR antibody and the goat-anti-rabbit antibody were used at dilutions of 1:1000. The rest of the Western blot protocols were performed as described [103, 106]. The band intensity was quantitated using digital image analyzer.

2.6 Transcription factor array

To detect transcriptional factors regulated by IL-17 stimulation in leukemia cells, we examined the profile of 54 transcriptional factors (TFs) using Panomics

Transcriptional Factor Array (1) Kit. Panomics TranSignal™ Protein/DNA Arrays simplifies the functional analysis of eukaryotic TFs and can be used to study TF activation in a variety of biological processes, including cell proliferation, differentiation, transformation, apoptosis and drug treatment [107]. The array membranes were spotted with 54 different consensus-binding sequences (oligos) and enable one to detect over 54 TFs at once in one treatment. Twenty million leukemia cells/ml were pretreated with vanadate (5 mM) for 30 minutes to inhibit endogenous phosphatases. Next, 5 million of the cells we either untreated or stimulated with IL-17 (100 ng/ml) for 4 hours in the incubator. The cells were packed by centrifugation at 1,800 x g for 3 minutes, washed 3x with PBS and gently lysed in a lysing buffer (see Western blot protocols above) without detergents by repeated aspiration through a 22-gauge needle to prevent rupturing of the nuclei. Intact nuclei were isolated by layering the cell lysate over 50% glycerol solution in Eppendorf tubes followed by centrifugation at 1,000 x g for 5 minutes. The supernatants were carefully aspirated, the nuclei pellet harvested and washed 2x with PBS. Next, the nuclei were disrupted in a nuclei lysing buffer (provided by the kit) and protein concentration determined as described above. In a slightly modified version, 12 ug of nuclei proteins in 200 ul of incubation buffer were incubated with each array membrane containing the oligos for hybridization of each oligo to its specific transcriptional factor in the nuclei extract. The membranes were washed several times and the oligo/transcriptional factor complexes (DNA/protein complexes) were detected by detection per the kit. The spots representing the various transcription factors were identified based on the charts provided by the kit. The intensities of the spots were scanned by digital image analyzer and the fold stimulation by IL-17 compared to the intensities in the untreated cells.

2.7 NF-kB/DNA and STAT3/DNA binding assays for detection of NF-kB and STAT3 activation by IL-17

To study the effect of IL-17 on NF-kB and STAT3 DNA binding functions [108, 109], 4 million leukemia cells/ml were untreated or IL-17 in time course experiments. The cells were used for specific NF-kB/DNA binding or STAT3/DNA binding assays using the NF-kB and STAT transcription factor assay kits (Active Motif, Chemicon). The kit enabled us to monitor the activation or repression of NF-kB or STAT3 proteins. The experiments were performed in triplicate.

2.8 Cell proliferation assays

To determine whether IL-17 stimulates cell proliferation in leukemia cells, 6 x 10^5 cells were either untreated or stimulated with IL-17 (100 ng/ml) for 48 hours [110]. The cells were harvested and aliquots were diluted into 0.4% trypan blue/PBS solution at a ratio of 1:10. The cells were counted in triplicate and the average viable cell count was recorded from each sample. We also performed MTT proliferation assay using 4 x 10^5 cells untreated or cells treated with IL-17 (100 ug/ml) for 48 hours in 96 well plates in triplicate. The rest of the details of the MTT assay protocols were as previously described [110].

2.9 Caspase3 activity assays as evidence for apoptosis

We purchased caspase3 assay kit from MBL International, Woburn, MA, US [111]. Sodium butyrate (NaB) is a strong inducer of apoptosis in cancer cells [112]. To determine the effects of IL-17 on NaB-induced apoptosis in leukemia cells we performed caspases3 enzymatic (colorimetric) assays in lysates from 15 x 10^6

untreated or NaB treated cells. Specifically, the cells were either untreated or treated with NaB (5 mM) alone, or treated with NaB (5 mM) plus IL-17 (100 ng/ml) or with IL-17 (100 ng/ml) alone in tissue culture media for up to 48 hours. At the end of the incubation, aliquot of cells were counted prior to lysing in RIPA buffer (provided in the caspase3 assay kit) and total lysate protein concentration was determined as indicated above. To measure caspase3 activity in the cell lysates, 30 ug total lysate protein from each sample was added to each experimental assay well. The enzymatic activity was measured in triplicate in microtiter plate reader according to instructions provided by the kit.

2.10 Assessment of chemotactic effects of media from IL-17 treated cells

In order to ascertain if media from IL-17 treated cells has chemotactic effects towards monocytes [103, 113], we employed the Boden Chamber chemotaxis assay method as previously described [103]. Specifically, 2000 monocytes/ml in fresh complete media were placed in the upper portion of the Boden Chamber. Equal volume of media from either untreated or IL-17 stimulated cells was put in the lower chamber to serve as a source of chemotaxis. The setup was incubated for 2-hours. Next, estimation of number of monocytes which crossed the membrane barrier to the lower chamber was performed according to the kit. Experiments were conducted in triplicate.

2.11 Assessing IL-17-induced cytokine expression by cytokine ELISA

In order to validate IL-17-induced cytokine expression observed in the cytokine antibody array, we performed cytokine ELISA assay as specified in the cytokine ELISA Kits (Ray Biotech, Norcross, GA) using equal amount of tissue culture proteins from untreated or IL-17 treated cells [103]. The ELISA monitored expression of IL-2, IL-3 and IL-8. In addition, to determine whether IL-17-induced IL-2 expression was mediated by either PI-3 K/Akt or by Jak2, in some experiments, we pre-incubated some of the cells with PI-3 K inhibitor LY20094 (20 nM) or Jak2 inhibitor AG490 (15 nM) for 30 minutes prior to stimulating the cells with IL-17 (100 ng/ml) for 24 hours.

3. Results

3.1 Effects of IL-17 on cytokine expression in human leukemia cells

To determine whether IL-17 stimulates cytokine expression in leukemia cells, we performed cytokine antibody array using tissue culture media from untreated and treated cell. As seen in **Figure 1**, within 24 hours IL-17 stimulated several fold differential expression of various cytokines in the ranking order of IL-2 > IL-3 > GRO > IL-10 > RANTES>IL-15 > IL-1. However, IL-17 failed to simulate IL-8 expression. Stimulation of cytokine expression by IL-17 was significantly inhibited by the PI-3 K inhibitor LY20094 (**Figure 1**), suggesting a role for PI-3 K in the mechanism by which IL-17 stimulates cytokine expression. Similar results were observed in THP-1 human leukemia cell line (data not shown). Also, a neutralizing antibody against of IL-17 blocked IL-17 from stimulating cytokine expression (data not shown), confirming that the observed stimulation of cytokine expression is attributed to L-17. Using ELISA assay, we confirmed stimulation of IL-2 and IL-3 expression by IL-17 without effect on IL-8 expression (**Figure 2**). The lack of effect of IL-17 on IL-8 expression seen in leukemia cells are in contrast to IL-17-induced

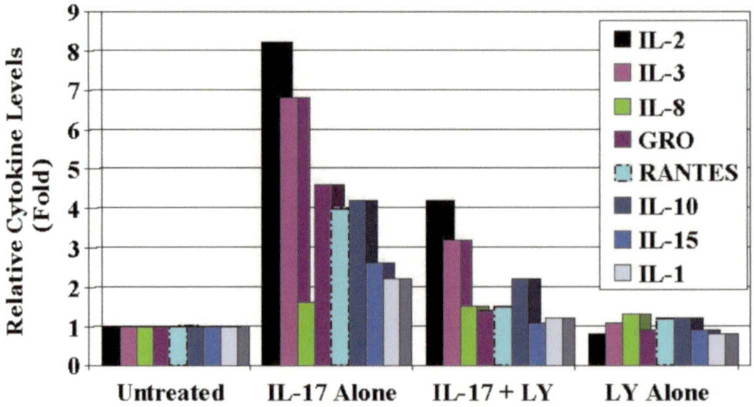

Figure 1.
IL-17 Stimulation of differential expression of cytokines: Inhibition by PI-3K Inhibitor (LY20094). Tissue culture media from untreated and treated cells were assayed for cytokine expression by cytokine-antibody array. Data is an average of two experiments.

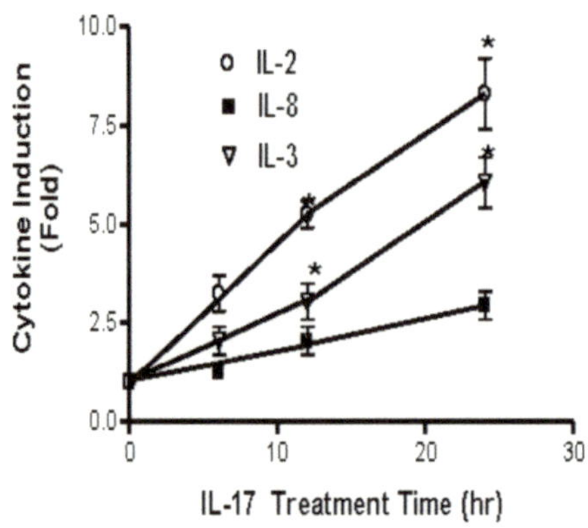

Figure 2.
Effects of IL-17 on IL-2, IL-3 and IL-8 expression. Asterisk (*) indicates significant differences between IL-17 treated and untreated cells.

IL-8 expression reported in different cell types [16–20]. As a follow up to our results in **Figure 1**, and our previous report that IL-17 activates Jak/STAT pathway [93], we determined whether both the PI-3 K and Jak2 mediate stimulation of specific cytokine expression by IL-17. To do so, we examined the effects of PI-3 K inhibitors LY20094 (LY) and wortmannin (WM) and Jak2 inhibitor AG490 (AG)) on IL-17-induced IL-2 expression. The ELISA array results in **Figure 3** indicate that individually LY20094 and wortmannin partially inhibited IL-17 stimulated IL-2 expression. The Jak2 inhibitor AG490 also exhibited similar inhibitory effect on ability of IL-17 to stimulate IL-2 expression. A combination of both LY20094 and AG490 completely blocked stimulation of IL-2 expression by IL-17 (not shown). These results confirmed roles for both PI-3 K and Jak2 in the mechanisms by which IL-17 stimulates IL-2 expression.

3.2 Media from IL-17 treated leukemia cells produce chemotaxis

Given that IL-17 stimulated significant expression of two chemokines (GRO and RANTES), we examined whether the culture media from IL-17 treated leukemia cells could serve as chemoattractant to human monocytes from PBMC. As seen in

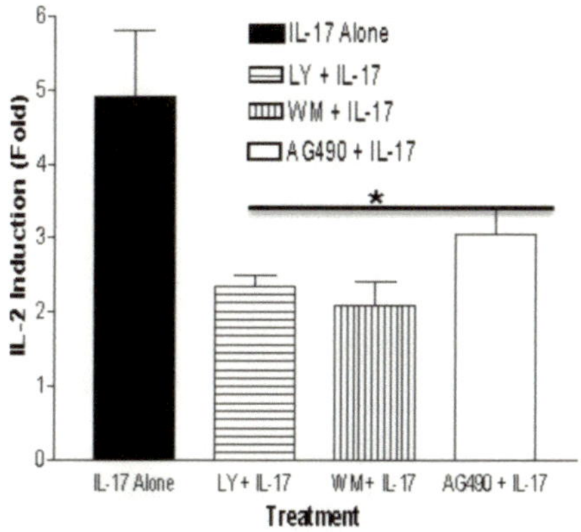

Figure 3.
Inhibition of IL-17-induced IL-2 expression by PI-3K inhibitors (LY and WM) and Jak2 inhibitor (AG). Asterisk (*) indicates significant differences between IL-17 alone and IL-17 plus PI-3K Inhibitors (LY and WM) or Jak2 inhibitor (AG490).

Figure 4, as compared to culture media from untreated cells, culture media from IL-17 treated cells exhibited significant time-dependent chemotaxis towards monocytes, confirming that culture media from IL-17 treated cells contains secreted chemotactic chemokines that induced chemotaxis [46].

3.3 IL-17 stimulates leukemia cell growth and protection from apoptosis

Next, we investigated the effect of IL-17 on leukemia cell proliferation. Untreated or IL-17 stimulated cells were assessed for cell growth using trypan blue exclusion and MTT assays. As shown in **Table 1**, IL-17 exhibited a time-dependent stimulation of leukemia cell growth by 3.3-fold within 48 hours. Similar results were seen in MMT assays (data not shown). Next, we examined whether IL-17 promotes leukemia cell survival and anti-apoptotic effects in leukemia cells by protecting the leukemia cells from apoptosis. The results in **Table 2** indicates that NaB alone causes significant reduction in leukemia cell survival from 100% to 52% in 24 hours. However, in the presence of IL-17, Na-induced decline in cell survival is

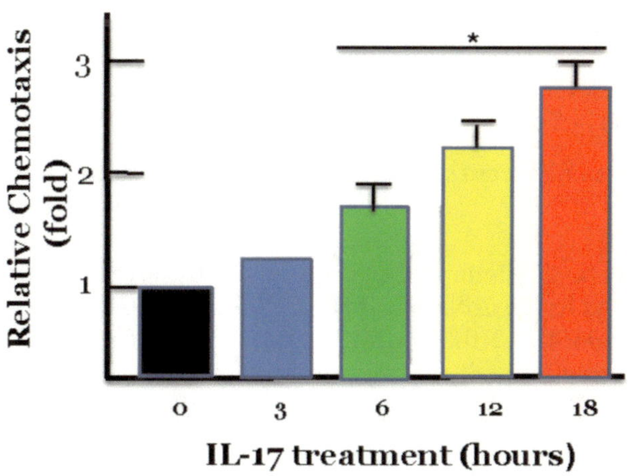

Figure 4.
IL-17 induced Chemotaxis. Tissue culture media from untreated IL-17 cells were assayed for chemotaxis in a Boyden Chamber with monocytes on the upper chamber.

Treatment	24 hours Cell Growth (Fold)	48 hours Cell Growth (Fold)
Cells	1.0	1.0
Cells + IL-17	2.3.0	3.3 ± 0.2

Table 1.
Effects of IL-17 on cell growth, cell survival and caspase3 activity—IL-17 promotes cell growth.

Treatment	24 hours Cell Survival (%)
Cells alone	100
Cells + NaB	52 ± 2.2
Cells + IL-17 + NaB	83 ± 1.3

Table 2.
Effects of IL-17 on cell growth, cell survival and caspase3 activity—IL-17 protects cells from butyrate-induced apoptosis.

Treatment	24 hour; Relative Caspase 3 Activity (Fold)	48 hours Relative Caspase 3 Activity (Fold)
Cells alone	1.0	1.0
Cells + NaB	3.2 ± 0.4	4.3 ± 0.1
Cells + IL-17 + NaB	1.7 ± 0.2	2.0 ± 0.3

Table 3.
Effects of IL-17 on cell growth, cell survival and caspase3 activity—IL-17 inhibits butyrate-induced caspase3 activation. Data represent mean plus/minus SD.

markedly inhibited cell survival improved from 52% to 83%. As shown in **Table 3**, NaB alone stimulated activation of caspase3 activity from 1-fold in untreated cells to 3.2 fold in 24 hours and 4.3-fold in 48 hours, indicating NaB-induced apoptosis in cells in the absence of IL-17. However, in the presence of IL-17, NaB-induced caspase3 activation is markedly reduced from 3.2-fold to 1.7-fold in 24 hours and from 4.3-fold to 2.0-fold in 48 hours. IL-17 also upgrades Bcl2 in the presence of NaB (not shown). Thus, IL-17 protects leukemia cells from undergoing apoptosis and enhances their survival. The results suggest that IL-17 may be inducing inactivation of pro-apoptotic signals while partially restoring the anti-apoptotic protein Bcl-2 expression.

3.4 IL-17 stimulates differential activation of transcription factors in leukemia cells

Within 4 hours IL-17 stimulated significant and differential activation of several transcription factors in the order of c-Myb (5.5-fold) > EGR-1 (5.0-fold) > STAT3 (4.0-fold)> > Smad3/4 (3.4-fold) > SRE (3.0 fold>CDP (2.5-.fold). IL-17 failed to activate NF-kB. Using individual transcription factor/DNA binding assays, we confirmed that STAT3/DNA binding activity is significantly enhanced by IL-17 (**Figure 5**). In contrast, IL-17 did not stimulate NF-kB/DNA binding activity in these leukemia cells (**Figure 6**). Together these results show that IL-17 differentially activates several transcriptional factors associated with regulation of cell growth, cell differentiation and apoptosis but failed to stimulate NF-kB in these cells even though IL-17 is known to stimulate NF-kB in many cell types [50]. Of note, NF-kB

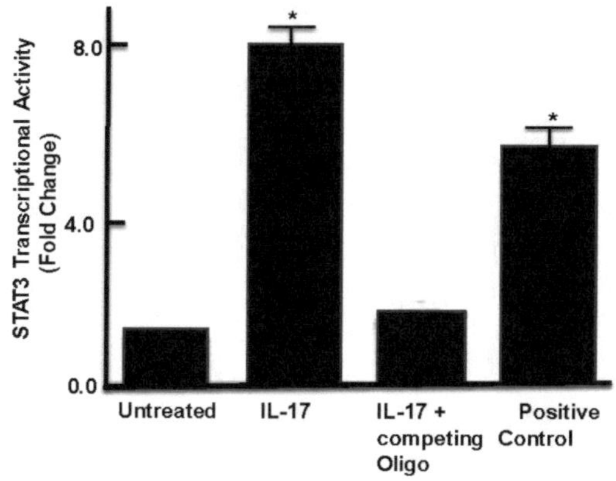

Figure 5.
Activation of STAT3 Transcriptional Activity by IL-17.

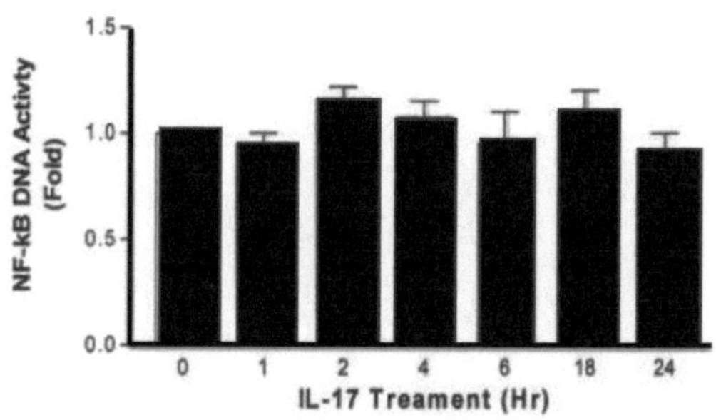

Figure 6.
Lack of IL-17 stimulation of NF-kB DNA Binding Activity.

is already highly constitutively expressed in active from in these leukemia cells. This could explain why IL-17 failed to stimulate NF-kB activation further in these cells.

3.5 Direct evidence that IL-17 activates PI-3 K/Akt signaling pathway in leukemia cells

Because the PI-3 K inhibitor Ly20094 inhibited IL-17-induced cytokine expression, we examined the direct effects of IL-17 on PI-3 K and Akt phosphorylation and activation. As shown in **Figure 7a** and **7b**, in as early as 0.5 minutes, IL-17 stimulated PI-3 K tyrosine phosphorylation by 4.5-fold. PI-3 K phosphorylation and activation usually lead to downstream Akt (PKB) activation [113]. Therefore, we next examined the effects of IL-17 on Akt (PKB) phosphorylation and activation. Akt can be phosphorylated on Serine 473 (Ser^{473}) and/or Threonine 308 (Thr^{308}), which is in the activation domain. The western blot results in **Figure 8a** show that IL-17 stimulated Akt phosphorylation on $Serine^{473}$ to 5-fold within 10 min in these cells. The results in **Figure 8b** show that IL-17 stimulates rapid phosphorylation of Akt on Thr^{308} with maximum effect noted at 5 minutes. Stimulation of Akt phosphorylation on $Serine^{473}$ by IL-17 was inhibited by the PI-3 K inhibitor wortmannin (WM) (no shown). These results imply that stimulation of Akt phosphorylation by IL-17 is mediated by PI-3 K. Once Akt is activated, it phosphorylates a host of downstream effectors including BAD, Caspase3, forkhead transcription factor (FKHR), glycogen synthase kinase-3 (GSK3-beta), AFX, eNOS,TSC2, MDM2, P21/CIP1 and other downstream effectors as shown in **Figure 9a** Dephosphorylated BAD, caspase3 and GSK3-beta play vital roles in induction of apoptosis [114].

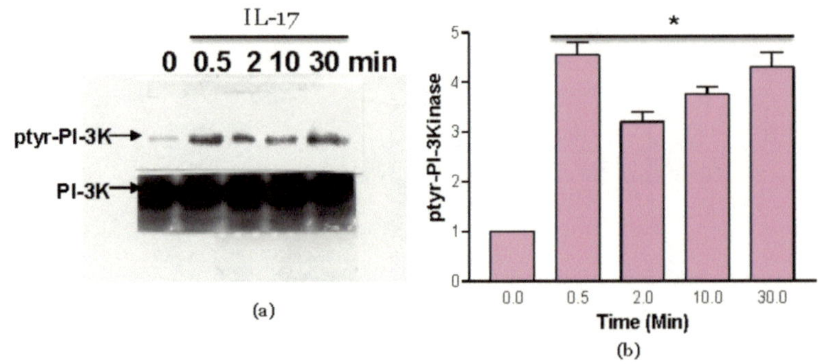

Figure 7.
Time course of IL-17-induced PI-3K phosphorylation detected by Western blot using either specific antibody for tyrosine phosphorylated PI-3K or total PI-3K as loading control. Scanned values represent ptyr-PI-3K/PI-3K ratios from 3 experiments (b). Asterisk (*) indicates significant differences between IL-17 treated and untreated cells. Results are representation form several experiments.

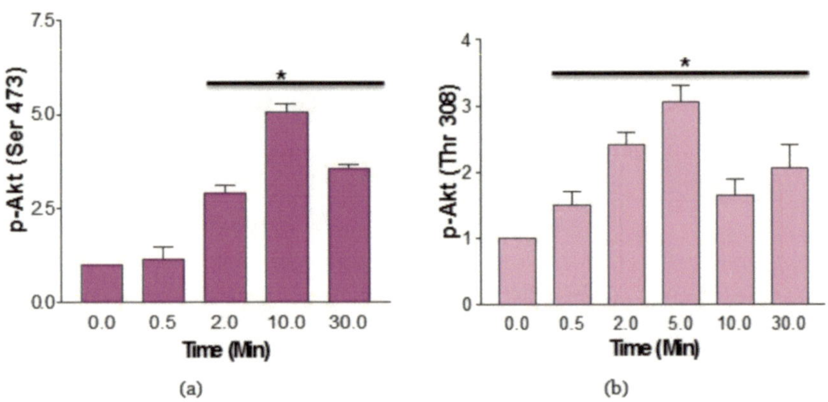

Figure 8.
Time course of IL-17 stimulation of Akt^{Ser473} phosphorylation (a) and (b) Akt^{Thr308} Specific antibody to either Akt^{Ser473} or Akt^{Thr308} was used to monitor Akt phosphorylation by Western blot. Blots were stripped and reprobed for total Akt for loading control. The blots from 3 experiments were scanned and results are presented. Asterisk (*) indicates significant differences between IL-17 treated and untreated cells.

However, upon their phosphorylation, these pro-apoptotic proteins lose their pro-apoptotic activities [114] as phosphorylation of both BAD, caspase3 and GSK-3 leads to their inactivation. The results in **Figure 9b** and **c** show that IL-17 stimulates Akt-mediated BAD, Caspase3 and GSK-3-beta phosphorylation as p-BAD,

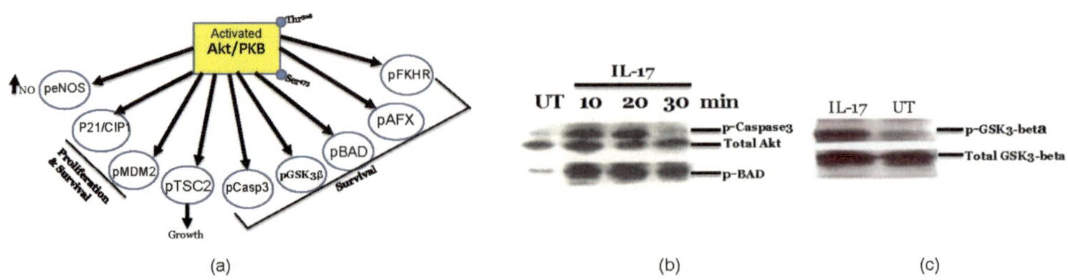

Figure 9.
Model showing activated Akt phosphorylation of its downstream targets (a). Effect of IL-17 on Caspase3 and BAD phosphorylation (b), GSK-3-beta phosphorylation (c). In (b) and (c) cells were untreated or stimulated with IL-17 and total cell lysates were monitored for Caspase3, BAD and GSK-3-beta phosphorylation by Western blot using specific phosphoprotein antibody to each protein. Total Akt (b) or total GSK-3-beta (c) was probed for loading control. Results are representation of several experiments.

p-Caspase3 and p-GSK3-beta were contained in Akt pulled down complex from IL-17 treated cells. Phosphorylation of caspse3, BAD, GSK3-beta and STAT3 are associated with enhanced cell survival [115, 116] and could explain in part how IL-17 promotes cell survival. Also, IL-17 stimulated Akt-dependent phosphorylation of mammalian target of rapamycin (mTor) on serine 2448 (motor $^{Ser\ 2448}$) [117], which was inhibited by the Akt inhibitor SH5 (not shown).

4. Discussions

We have provided strong evidence that IL-17 stimulates significance and differential expression of IL-2, IL-3, IL-10, IL-15, GRO, and RANTES in human leukemia cells. The stimulatory effect of IL-17 on cytokine expression in these cells is similar to previous reports in non-hematopoietic cells by IL-17 [16–20]. However, IL-17 does not stimulate IL-8 expression in these cells, which contradicts early reports that IL-17-induces IL-8 expression in different cell types [17–20]. Induction of cytokine expression by IL-17 in these leukemia cells could have strong biological relevance in vivo because increases in IL-17 level in a tumor microenvironment can trigger induction of other cytokines including chemokines that could generate combination of proinflammatory, anti-inflammatory and chemotactic responses [16–20, 54]; [79–81]. IL-2 is a proinflammatory cytokine [118], which also regulates helper T cell differentiation [119]. IL-3 stimulates regulation of multipotent hematopoietic stem cell function and differentiation of all lineages as well as promote proliferation of myeloid progenitor cells [120, 121]. IL-10 is a master regulator of immunity to infection and an anti-inflammatory cytokine that can counteract the pro-inflammatory effects of IL-2 [122]. Secondly, IL-10 is known to synergize with IL-2 to promote CD8 $^+$ T cell cytotoxicity [123]. IL-15 is known to suppress apoptosis in T-lymphocytes by inducing Bcl2 and/or Bcl-xl in humans [124]. Perhaps, IL-15 contributes to the anti-apoptotic effect of IL-17 in these cells. Both IL-2 and IL-15 have structural and functional similarities, share the common gamma chain of their receptors and promote immune response [125]. Both GRO and RANTES are chemokines and are associated with induction of chemotaxis and recruitment of neutrophils and macrophages to sites of infection [37]. Thus, IL-17- induced GRO and RANTES expression and secretion from leukemia cells into the culture media, could account for the chemotactic effect IL-17 seen in our studies.

These leukemia cells express receptors for some of the cytokines secreted to the culture media in response to IL-17. Therefore, some of the cytokines secreted into culture media can promote both autocrine and paracrine effects on the leukemia cells. We have provided evidence that IL-17 stimulates phosphorylation of the pro-apoptotic proteins BAD, caspase3 and GSK3-beta, thus negating their functions. In addition, IL-17 promotes inhibition of Caspase3 activity in these leukemia cells. Furthermore, IL-17 enhances Akt phosphorylation and activation, which are associated with cell survival [126]. The ability of IL-17 to enhance protection of the leukemia cells from apoptosis implies that elevated IL-17 levels in a tumor microenvironment could lead to promotion of leukemia cell proliferation and survival, both of which could potentially produce poor prognosis in leukemia patients. Another interesting outcome of this study is that IL-17 stimulates activation of several transcriptional factors including cMyb, EGR-1, STAT3, Smad3/4, SRE, CDP, which are known to regulate proliferation, differentiation and survival [115, 127, 128]. This effect of IL-17 could in part contribute to the mechanism growth promotion and survival in these leukemia cells. Stimulation of smad3/4 transcriptional factors of the TGF-beta signaling pathway [102, 115] by IL-17 may point to potential cross talk between IL-17 and TGF-beta-induced signaling pathways to synergize their biological effects [127, 128]. The lack of activation of NF-kB by IL-17 in these leukemia cells is not surprising since typically these leukemia cells constitutively express high levels of active NF-kB, which could explain the apparent lack of NF-kB response to IL-17. Lack of NF-kB activation by IL-17 in these cells is in contrast to IL-17-induced NF-kB activation reported in many cells [50].

We have provided ample evidence that IL-17 activates and utilizes the Jak/STAT signaling pathway in these leukemia cells. In this pathway, IL-17 stimulates phosphorylation of Jak1, Jak2 and Jak3, STAT1, STAT2, and STAT3 [55, 93, 97]. We have also shown that Jak2 partially mediates IL-17-induced IL-2 expression. Furthermore, IL-17 strongly stimulates phosphorylation and activation of PI-3 K/ Akt pathway and promoting Akt-mediated phosphorylation of its downstream effectors. Another interesting observation is that Akt-partially mediates stimulation of IL-2 expression and secretion by IL-17. Also, IL-17 promotes phospho-Akt's association with BAD, caspase3 and GSK3-beta, supporting Akt-mediated phosphorylation of these proteins in IL-17 treated cells. These observations could in part explain how IL-17 promotes anti-apoptosis and survival in these leukemia cells [75–77, 126].

IL-17 stimulates activation of Raf–MEK–ERK1/2 pathway [92–95, 101], which could partially account for the growth promoting effects of IL-17 in leukemia cells. Previous thesis research in our laboratory revealed that IL-17 stimulates activation of LCK [129] and PKC [130]. Also, IL-17 promotes association between LCK and the p85 subunit of the PI-3 K, thus providing another mechanism for PI-3 K activation by IL-17 via LCK, a member of the Src kinases family [129]. Activation of PKC by IL-17 is associated with enhanced PKC ability to regulate cell cycle progression in leukemia cells [130]. As indicated earlier, IL-17 is profoundly implicated in many human diseases [55–74], thus supporting the suggestion that design and production of anti-IL-17 drugs could lead to better strategies for development of new therapies for those diseases [88–90]. Although the recent reports implicating IL-17 in the mechanism of the "cytokine storm" event in COVID-19 infection is far from

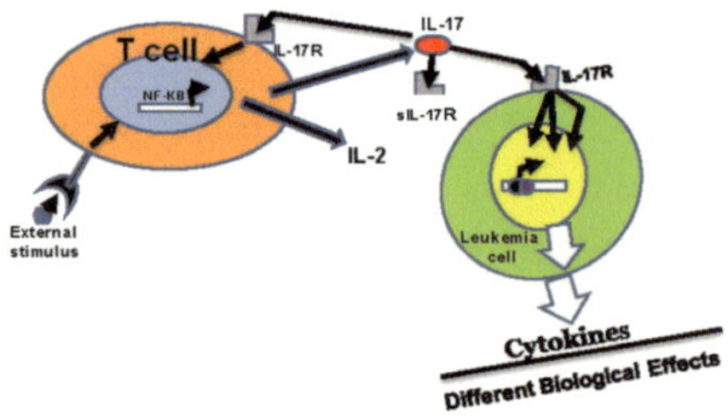

Figure 10.
Model showing activated memory T cell secreted IL-17: Paracrine mechanism of how secreted IL-17 activates cytokine expression and secretion in leukemia cells.

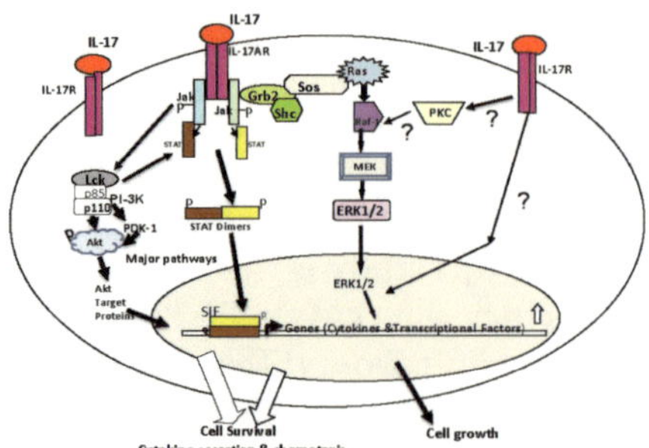

Figure 11.
Model showing multiple signaling mechanisms used by IL-17 in Leukemia Cells.

conclusion, there are calls for development of anti-IL-17 drugs as adjunct therapy for diseases in which IL-17 plays an active role [87, 131]. IL-17-enhanced leukemia cell growth, survival and anti-apoptosis strengthens the argument in favor of inclusion of leukemia in the list of human diseases for which anti-IL-17 adjunct therapy should be considered. Our model in **Figure 10** explains the paracrine role of T-cell secreted IL-17 in leukemia cells. Elucidation of the multiple signaling mechanisms of IL-17 in leukemia cells in our study and illustrated in **Figure 11** further enrich our knowledge on the biological effects and mechanisms of IL-17.

5. Conclusion

Our studies on effects and mechanisms of IL-17 in human U937 leukemia cells revealed that these cells express IL-17A receptor and IL-17 stimulates cell growth, survival, chemotaxis and differential expression of cytokines. These results suggest that IL-17 could trigger expression and secretion of various cytokines including chemotactic chemokines in leukemia patients. Also, IL-17 promotes anti-apoptotic effects in these cells. If these biological effects of IL-17 described here, were to occur in leukemia patients, IL-17 could promote poor prognosis in the patients. Furthermore, IL-17 stimulates differential activation of several transcriptional factors including c-Myb, EGR-1, STAT3, smad3/4 CDP and SRE but not NF-kB in these cells. Lastly, multiple signaling pathways including PI-3 K/Akt, Jak/STAT, Raf–MEK-ERK-1/2 and Lck signaling pathways differentially mediate the biological effects of IL-17 in the U937 leukemia cells. Any of these pathways could serve as a target for anti-IL-17 drugs.

Acknowledgements

This work was partially supported by NIAMS/NIH R03 grant, U54 cancer partnership NCI grant U54CA091408 and NIGMS/NIH SCORE grant to Professor Adunyah, who was also supported by cancer partnership grant U54CA163069/NCI during preparation of this chapter. Professor Arthur was partially supported by Biochemistry Department, KNUST, Kumasi, Ghana during his sabbatical. We thank Dr. S. V. Subramaniam and W. Williams for their contribution in the initial stages of this work.

Author details

Samuel Evans Adunyah*, Richard Akomeah, Fareed K.N. Arthur, Roland S. Cooper and Joshua C.M. Williams
Department of Biochemistry, Cancer Biology, Neuroscience and Pharmacology, Meharry Medical College, Nashville, TN, USA

*Address all correspondence to: sadunyah@mmc.edu

References

[1] Rouvier E, Luciani MF, Mattéi MG, Denizot F, Golstein P. CLTA cloned from activated T cell, bearing au-rich messenger RNA instability sequences, and homologous to a herpes saimiri gene. The Journal of Immunology. 1993; 150(12):5445–5456

[2] Yao Z, Painter SL, Fanslow WC, Ulrich D, Macduff BM, Springs MK, Armitage RJ. Human IL-17: a novel cytokine derived from T cells. J. Immunology.1995b;155:5483–5486

[3] Avarvak T, Chabbbaud M, Natvig JB. IL-17 produced by some pro-inflammatory Th1/Th0 cells but not Th2 cells. J. Immunol.1999;162:1246–1251

[4] Tartour E, Fossiez F, Joyeux I, Galinha A, Gey A, Claret E, et al. Interleukin-17, a T-cell-derived cytokine, promotes tumorigenicity of human cervical tumors in nude mice. Cancer Research. 1999;59:3698–3704

[5] Albanesi C, Caavni A, Giromori G. IL-17 is produced by nickel-specific T lymphocytes and regulates ICAM-I expression and chemokine production in human keratinocytes: synergistic or antagonist effects with IFN-gamma and TNF-alpha. J. Immunol. 1999;162: 494–502

[6] Shin HCK, Benbemou N, Esnault S, Guenounou M. Expression of IL-17 in human memory CD45RO+T lymphocytes and its regulation by protein kinase A. 1999;11(4):257–266.

[7] Kotake S, Udagawa N, Takahashi N, et al. IL-17 in synovial fluids from patients with rheumatoid arthritis is a potent stimulator of osteoclastogenesis. J. Clin. Invest. 1999;103(9):1345–1352

[8] Nakae S, Saijo S, Horai E, Sudo K, Morai S, Iwakura Y. IL-17 production from activated T cells is required for the spontaneous development of destructive arthritis in mice with deficient in IL-1 receptor antagonist. Proc. Nat. Acad. Sci. 2003;100(10):5986–5990

[9] Betelli E, Korn T, Kuchroo VK. Th17: The third member of the effector T cell Trilogy. Curr. Opin. Immunol. 2007;19 (6):652–657

[10] Chen Z, Osahea JJ. Th17 Cells: a new fate for differentiating helper T cells. Immunol. Res. 2008;41:87–102

[11] Korn T, Bettelli E, Oukka M, Kuchroo VK. IL-17 and Th17 cells. Ann. Rev. Immunol. 2009;27:495–517

[12] Lenarczyk A, Helsloot J, Farmer K, Peters L, Sturgess A, Kirkham B. Antigen-induced IL-17 response in peripheral blood mononuclear cells PBMC) of healthy controls. Clin. Exp. Immunol. 2000;122(1):41–48

[13] Kuiper, JJW, Emmelot ME, Rothova A, Mutis T. Interleukekin-17 production and T helper cells in peripheral blood mononuclear cells in response to ocular lysate in patients with birdshot chorioretinopathy. Molecular Vision. 2013;19:2606–2614

[14] Lourda M, Olsson-Åkefeldt S, Gavhed D, et al. Detection of IL-17A-producing peripheral blood monocytes in Langerhans cells histiocytosis patients. Clin. Immunol. 2014;153(1) 112–122

[15] Szabo PA, Goswami A, Mazzuca DM, Kim K, et al. Rapid and Rigorous IL-17A production in distinct subpopulation of effector memory T lymphocytes constitute a novel mechanism of toxic shock syndrome immunopathology. Journal Immunol.2017;doi:10.4049/ jimmunol.1601366:1-19

[16] Hirata T, Osuga Y, Hamasaki K, Yoshino O, Ito M, Hasegawa A. et al.

Interleukin-17 stimulates IL-8 secretion, cyclo0xygenease-2 expression, and cell proliferation of endometriotic stromal cells. Endocrinology. 2008;149(3): 1260–1267

[17] Kuwabara T, Ishikawa F, Kondo M, Kakiuchi T. The role of IL-17 and related cytokines in Inflammatory autoimmune diseases. Mediators of Inflammation. 2017;2017:doi.org/10.1155/2017/390861, 11 pages

[18] Numasaki M, Tomioka Y, Takahashi H, Sasaki *H. IL*-17 and IL-17F modulate GM-CSF production by lung microvascular endothelial cells stimulated with IL-1beta and/or TNF-alpha. Immunology Letters, 2004;95(2): 175–184

[19] Fossiez F, Djossou O, Chomarat P, Flores-Romo L, Ait-Yahia S, Maat C, et al cell Interleukin-17 induces stromal cells to produce proinflammatory and hematopoietic cytokines. J. Exp. Med. 1996;183(6):2593–2603

[20] Honda K, Wada H, Nakamura M, Nakamoto K, et al. IL-17A synergistically stimulates TNF-alpha-induced IL-8 production in human airway epithelial cells: A potential role in amplifying airway inflammation. Exp. Lung Res. 2016;42(4):2052016

[21] Yao Z, Springs MK, Deny JMJ, Stockbine L, Park LS, VendenBos T, Zappone J, Painter SL, Armitage RJ. Molecular Characterization of the human IL-17 receptor. Cytokine. 1997;9 (11):794–800

[22] Gaffen S. Structure and signaling of the IL-17 receptor superfamily. Nat Rev. Immunol. 2009;9(8):556].

[23] Honrati MC, Meliconi R, Pulsatelli S, Cané S, Frizziero L, Facchini. A High in vivo expression of Interleukin-17 receptor in synovial endothelial cells and chondrocytes from arthritis patients. Rheumatology. 2001; 40(5):522–527

[24] Haudenchild D, Moseley T, Rose L, Redd AH. Soluble and transmembrane isoforms of novel Interleukin-17 receptor-like protein by RNA splicing and expression in prostate cancer. J.Biol. Chem. 2002;277(6):4309–4316

[25] Kawaguchi M, Adachi M, et al. IL-17 cytokine Family. J. Allergy and Clin, Immunol. 2004;114(6):1265–1273.

[26] Gaffen SL, Kramer JM, Yu JJ, Shen F. The IL-17 Cytokine Family. Vitamins and Hormones. 2006;74:255–282

[27] Li H, Chen J, Huang A, Stinson J, Heldens S, Foster J, Dowd P, Gurney A, Wood WL. Cloning and characterization of IL-17B And IL-7C, two new members of the IL-17 family. Proc. Nat. Acad. Sci. 2000;97(2):773–778

[28] Lee J, Maruoka M, Corpuz T, Baldin DT, Foster JS, Goddard AD, Yamsura DG, Vandlen RL, Gurney AL. IL-17E, a novel proinflammatory ligand for the IL-17 receptor homolog IL-17Rh1.J. Biol. Chem. 2001;276(2): 1660–1664

[29] Moseley A, Haudenchild DR, Rose L, Reddi AH. Interleukin-17 family and IL-17 receptors. Cytokine Growth Factor Rev. 2003;14(2):155–174

[30] McAllister F, Henry A, Kreinder JJ, et al. Role of IL-17A, IL-17F, and the IL-17 receptor in regulating stimulating factors in bronchial epithelial: implications for airway inflammation in Cystic Fibrosis. J. Immunol. 2005;175 (1):404–412

[31] Kuestner RE, Taft DW, Haran A, Brandt CS, Brender T, Lum K, et al. Identification of the IL-17 receptor related molecule IL-17C as the receptor for IL-17F. J. Immunol. 2007;179(8): 5462–5373

[32] Han Q, Das S, Hirano M, Holland SJ, McCurley MD, Guo P, Rosenberg CS, Boehm T, Cooper MD. Characterization of Lamprey IL-17 Family members and their receptors. J. Immunol. 2015;195: 5440–5451

[33] Broxmeyer HE, Starnes T, Ramey H, Cooper S, Dahl R, Williamson E, Hromas R. The IL-17 cytokine family members are inhibitors of human hematopoietic progenitor proliferation. Blood. 2006,108(2):770–771

[34] Krstic A, Mojsiović S, Bugarski D, The potential of Interleukin-17 to mediate hematopoietic response. Immunologic Res. 2012;52:34–41

[35] Mojsilović S, Jauković A, Santibaňez JF, Bugarski, D. Interleu-17 and its implication in the regulation of differentiation and function of hematopoietic and mesenchymal stem cells. Mediators of Inflammation. 2015; doi.org/10.1155/2015/470458

[36] Schwarzenberger P, La Russa V, Miller A, Ye P, Huang W, Zieske A, Nelson S, et al. IL-17 stimulates granulopoiesis in mice: use of an alternative, novel gene therapy-derived method for in vivo evaluation of cytokine. J Immunol. 1998;161:6383–6389

[37] Witowski J, Pawlaczyk K, Breborowicz A, Scheuren A, Kuzlan-Pawlaczk, Wiseniewska J, Polubinska A, et al. IL-17 stimulates intraperitoneal neutrophils infiltration through GRO-alpha chemokine from mesothelial cells. J. Immunol. 2000;165:5814–5821

[38] Schwarzenberger P, Huang W, Oliver P, Byme P, La Russa V, Zhang Z, Kolls JK. IL-17 mobilizes peripheral blood stem cells with short-and long-term responding ability in mice. J. Immunol. 2001;167:2081–2086

[39] Bradley Fellow S, Schurr JR, Kolls JK, Bagby GJ, Schwarzenberger P, Ley K. Increased granulopoiesis through IL-17 and granulocyte colony-stimulating factor in leukocyte adhesion molecule-deficient mice. Blood. 2001; 98:3309–3314

[40] Huang H, Kim HJ, Chang EJ, Lee ZH, Huang SJ. IL-17 stimulates the proliferation and differentiation of human mesenchymal stem cells. Cell Differ. 2009;16(10):1332–1343

[41] Han X, Yang Q, Shi Y. Interleukin-17 enhances immunosuppression by mesenchymal stem cells. Cell Death and Differ. 2014;21:1758–1768

[42] Ma T, Wang X, Jiao Y, Halto W, Qi Y, Gong H, Zhang L, Jiang D. Interleukin-17-induced mesenchymal stem cells prolong the survival of allogeneic skin. Ann Transplant. 2018; 23:615–621

[43] Liao S, Zhang C, Jin L, Yang Y. IL-17 alters the mesenchymal stem cell niche towards osteogenesis in corporation with osteocytes. J. Cell. Phys. 2020;235 (5):4466–4480.

[44] Laan M, Prause O, Miyamoto M, Sjostrand M, et al. A role of GM-CSF in the accumulation of neutrophils in the airways caused by IL-17 and TNF. Euro. Res. J. 2003;21(3):387–393

[45] Schwarzenberger P, Huang W, Ye P, Oliver P. Requirement of endogenous stem cell factor and Granulocyte-Colony Stimulating Factor for IL-17-mediated granulopoiesis. The J. Immunol. 2000;164(9):4783–4789

[46] Shahara S, Pickins SR, Mandelin II AM, Karpus, WJ, Huang Q, Kolls JK, Pope RM. IL-17-mediate monocyte migration occurs partially through CCL2/MCP-1 induction. J. Immunol. 2010;184(8):4479–4487

[47] Chin C-C, Chen C-N, Kuo H-C, Shi C-S, Hsieh MC, et a. Interleukin-17

induces CC chemokine receptor 6 expression and cell migration in colorectal cancer cells. J. Cell Physiol. 2015;230(7):1403–1437

[48] Chung AS, Wu X, Zhuang G, et.al. An interleukin-17-mediated paracrine network promotes tumor resistance to anti-angiogenic therapy. Nature Med.2013;19:1114–1123 Res. Ther. 2004; 6(2).doi.org/10.1186/ar1038

[49] Wang L, Yi T, Kortylewski M, Pandoll DM, Zang D, Yu *H. IL*-17 can promote tumor growth through and IL-6-STAT3 signaling pathway. J. Exp. Med. 2009;206(7):1457–1464

[50] Hwang S-Y, Kim J-Y, Kim K-W, ParkM-K, Moon Y, Kim W-U, Kim H-Y. IL-17 induces production of IL-6 and IL-8 in rheumatoid arthritis synovial fibroblasts via NF-kB- and PK-3Kinase/Akt-dependent pathways. Arthritis Res. & Therapy. 2004;6(2):1–9

[51] Zaretsky M, Etzyoni R, Kaye J, Sklair-Tarvon L, Aharoni A. Directed for evolution of a soluble human IL-17A receptor for the inhibition of psoriasis plaque formation in a mouse model. J. Chem. Biol. 2012. Doi.org/10.1016/J. ChemBiol.2012.11.012

[52] Gaffen SL. Role of IL-17 in the pathogenesis of Rheumatoid arthritis. Curr. Rheumatol. Rep. 2009;11(5): 385–370

[53] Robert M, Miossec P. IL-17 in rheumatoid arthritis and precision medicine: from synovitis expression to circulating bioactive levels. Frontiers of Med. 2018;doi.org/10.3389/fmed .2018.00364

[54] Beringer A, Miossec P. Systemic effects of IL-17 in inflammatory arthritis. Nat. Rev. Rheumatol. 2019:15: 491–501

[55] Krejsgaard T, Ralfkiaer U, Clasen-Linde E, Eriksen KW, at al. Malignant cutaneous T-cell lymphoma cells express IL-17 utilizing the Jak3/Stat3 signaling pathway. J. Invest. Dermatol. 2001;131(6):1331–1338

[56] Ferrarini I, Rigo A, Zamb A, Vinante F. Classical Hodgkin lymphoma cells may promote an IL-17-enriched microenvironment. Leukemia and Lymphoma. 2019;60(14):1–11

[57] Zhong W, Li Q. Rituximab or irradiation promotes IL-17 secretion thereby induces resistance to rituximab or irradiation. Cell. Mol. Immunol. 2017; 14:1020–1022

[58] Prabhala RH, Pelluru D, Fulciniti M, Song W, Pai C, et. Al. Elevated IL-17 level produced by Th-17 cells promote myeloma cell growth and inhibits immune function in multiple myeloma. Blood. 2010;115(26):5385–5392

[59] Prabhala RH, Fulciniti M, Pelluru D, Rashid N, Nigroiu A, Nanpappa P, et al. Targeting IL-17A in multiple myeloma: a potential novel therapeutic approach in myeloma.Leukemia. 2016;30(2):379–389

[60] Calcinotto A, Brevi A. Chesi M, et al. Microbiota-driven IL-17-producing cells and eosinophils synergize accelerate multiple myeloma progression. Nature Commun. 2018;9: 4832.doi.org/10.1038/s41467-018-07305-8

[61] Yanagisawa H, Hashimoto H, Minagawa S, Takasaka N, Moermans MR, et al. Role of IL-17 in murine model of COPD airway disease. Am J. Physiol. Lung Cell Mol.Physiol. 2017;312(1):L122-L130.doi.10.1152/ ajplumg.00301.2016

[62] Hynes GM, Timothy SC, Hinks ER.J. The role of Interleukin-17 in asthma: a protective response?. Open Research. 2020;6:00364–2019;doi.10.1183/ 23120541.00384-2019

[63] Lee Y. The role of interleukin-17 in bone metabolism and inflammatory

skeletal diseases. BMB Rep.2013;46(10): 479–483

[64] Yago, T, Nanke Y, Ichikawa N, Kobashigawa T, Mogi M, Kamatani N, Kotake S. IL-17 induces osteoclastogenesis from human monocytes alone in the absence of osteoblasts, which is potentially inhibited by anti-TNF-alpha antibody: a novel mechanism of osteoclastogenesis by IL-17. J. Cell Biochem. 2009;108: 947–955

[65] Fujino, S, Andoh A, Bamba S, Ogawa A, Hata K, Anaki Y, Bamba T, Fujiyama Y. Increased expression of interleukin-17 in inflammatory bowel disease. Gut. 2003;52(1):65–70

[66] Kuwabara T, Ishikawa T, Kondo M, Kakiuchi T. The role of IL-17 and related cytokines in inflammatory autoimmune diseases. Mediators of Inflammation. 2017;2017.Doi.org/10.1155/2017/3908061

[67] Yang B, Kang H, Fung A, Zhao H, Wang T, Ma D. The role of interleukin-17 in tumor proliferation, angiogenesis and metastasis. Mediators of Inflam. 2014.doi.org/10.1155/2014/623739

[68] Zhao J, Chen X, Herjan T, Li X. The role of interleukin-17 in tumor development and progression. J. Exp. Med. 2019;217(1)1–13.

[69] Zhu L, Mulcahy A, Mohammed AAA, et al. IL-17 expression by breast -cancer-associated macrophages: IL-17 promotes invasiveness of breast cancer cell lines. Breast Cancer Res. 2008;10(6): article R95

[70] Bian G. Zhao W-Y. "IL-17, an important prognostic factor and potential therapeutic target for breast cancer"? Euro J. Immunol. 2014;33(2): 604–605

[71] Cochaud S, Giustiniani J, Thomas C, et al. IL-17A is produced by breast cancer TILs and promotes chemoresistance and proliferation through ERK1/2. Sci. Rep 3, article #3456 (2013).doi.org/10.1038/srep03456

[72] Tartour E, Fossiez F, Joyeux I, Galinha A, Gey A, et al. Interleukin-17, a T-cell-derived cytokine, promotes tumorigenicity of human cervical tumors in nude mice. Can. Res. 1999;59: 3698–3704

[73] Bonin CM, Almeida-Lugo LZ, dos Santo, AR, Junqueira CT, et al. Interleukin-17 expression in the serum and exfoliated cervical cells of patients with high-risk oncogenic human papillomavirus. Cytokine. 2019;120: 97–98

[74] Xiang T, Long H, He L, Han X, Lin K, Liang Z, Zhuo W, Xie R, Zhu B. Intterleukin-17 produced by tumor microenvironment promotes self-renewal of CD133$^+$ cancer stem-like cells in ovarian cancer. Oncogene. 2015;34 (2):165–176

[75] You Z, Shi X-B, DuRaine, G, Haudenschild D, Clifford G, et al. Interleukin-17 receptor-like gene is a novel anti-apoptotic gene highly expressed in androgen-independent prostate cancer. Can Res. 2006;66(1): 175–183

[76] Zhang Q, Liu S, Ge D, Zhang Q, Xue Y, Xiong Z, et al. Interleukin-17 promotes formation and growth of prostate cancer adenocarcinoma in mouse models. Can Res. 2012;72(10); 2589–2599

[77] Zhang Q, Liu S, Parakuli KR, Zhang W, Zhang K, Mao Z, et al. Interleukin-17 promotes prostate cancer via MMP-induced epithelial-to-mesenchymal transition. Oncogene. 2017;36(5):687–699

[78] Waisman A, Hauptmann J, Regen T. The role of IL-17 in CNS diseases. Acta Neuropathilogica. 2015;129:625–637

[79] Milovanovic J, Arsenijevic A, Stojanovic B, et al. Interleukin-17 in chronic inflammatory neurological diseases. Frontier in Immunol. 2020;11 (947):doi.10.3389/fimmu.2020.00947

[80] Parajuli P, Mittal S. The role of IL-17 Giloma. J Spine and Neursurgey. 2013; suppl.1:S1–004.doi.10.4172/2325-9701. s1-004

[81] Fietcher JM, Moran B, Petrasca A, Smith CM. IL-17 in inflammatory skin diseases psoriasis and hidradenitis suppurativa. Clin. Exp. Immunol. 2020; 201(2):121–134

[82] Kaur R, Rawat KA, Kumar S, Aadi W, Akhtar T, Narang T, Dimple C. Association of genetic polymorphism of IL-17A and IL-17F with susceptibility of psoriasis. India J. Med. Res. 2018;148(4): 422–426

[83] Blauvelt A, Chiricozzi A. The immunologic role of IL-17 in psoriasis and psoriatic arthritis pathogenesis. Clin. Rev. Allergy & Immunol. 2018;55 (3):379–390

[84] Wu D, Yang XO. Th-17 responses in cytokine storm of COVID-19: an emerging target of Jak2 inhibitor Fedratinib. J. Microbiol. Infect. 2020;53 (3)368–370

[85] Raucci F, Mansour AA, Casillo MG, Saviano A, et al. Interleukin-17 (IL-17A), a key molecule of innate and adaptive immunity and its potential involvement in COVID-19-related thrombotic and vascular mechanisms. Autoimun. 2020;19(7):102572.

[86] Shibabaw T. Inflammatory cytokine: IL-17A signaling pathway in patients with COVID-19 and current treatment strategy. J. Inflamm. Res. 2020;13:673–680

[87] Bulat V, Situm M, Azadajic MD, Likic R. Potential role of IL-17 blocking agents in treatment of COVID-19. Br. J. Clin. Pharmacol. 2020;Jul 5.doi.org/ 10.1111/bcp.14437

[88] Canavan TN, Elmets CA, Cantrell WL, Evans JM, Elewski BE. Anti-IL-17 medications used in the treatment of plaque psoriasis and psoriatic arthritis: A comprehensive review. Am. J. Clin. Dermatol. 2016;17 (1):33–47

[89] Yin Y, Wang M, Wu J. Efficacy and safety of IL-17 inhibitors for the treatment of ankylosing spondylitis: a systemic review and meta-analysis. Arthritis Res. & Therapy. 2020;22(111). Doi..org/10.1186/s13075-020-02208-w

[90] Pacha O, Sallman MA, Evans SE. COVID-19: a case for inhibiting IL-17? Nature Reviews Immunol. 2020;20: 345–346

[91] Shalom-Barak T, Quach J, Lotz M. Interleukin-17-induced gene expression in articular chondrocytes is associated with activation of mitogen-associated protein kinases and NF-kB. J. Biol. Chem. 1998;42(16):27457–27473

[92] Subramaniam SV, Pearson LT, Adunyah SE. Interleukin-17 induces rapid tyrosine phosphorylation and activation of raf-kinase in human monocytic progenitor cell line U937. Biochem. Biophys. Res. Commun. 1999; 259(1):172–177

[93] Subramaniam SV, Cooper RS, Adunyah SE. Evidence for the involvement of JAK/STAT pathway in the signaling mechanism of interleukin-17. Biochem. Biophys. Res. Commun. 1999;262:14–19

[94] Lann M, Lötvall J, Fan Chun K, Lindén A. IL-17-induces release in human bronchial epithelial cells in vitro: role of mitogen-activated protein (MAP) kinases. British J. Pharmacol. 2001;133:200–206

[95] Chen Y, Kijstra A, Chen Y, Peizeng Y. IL-17A stimulates the

production of inflammatory mediators via Erk1/2, p38, PI-3K/Akt and NF-kB pathways in ARPE-19 cells. Molecular Vision. 2011;17:3072–3077

[96] Zheng Q, Diao S, Wang Q, Zhu C, et al. IL-17 promotes cell migration and invasion of glioblastoma cells via activation of PI-3K/Akt signaling pathway. A J. Cell Mol. Med. 2019;23(1): 357–369

[97] You T, Yihui Bi, Li J, Zhang M, Chen X, Zhang K. IL-17 induces reactive astrocytes and up-regulation of vascular endothelial growth factor (VEGF) through JAK/STAT signaling. Sci. Rep.7.2017:41779 doi.org/10.1038/sre p41779

[98] Amatya N, Garg AV, Gaffen S. IL-17 Signaling: The Yin and the Yang. Trends in Immunol. 2017;39(5)310–322

[99] Swaidani S, Liu C, Zhao J, Bulek K, Li X. TRAF Regulation of IL-17 signaling. Front Immunol. 2019;doi.org/ 10.3389/fimmu.2019.0129

[100] Li X, Bechara R, Zhao J, McGeachy, MJ, Gaffen S. IL-17 receptor-based signaling and implications for diseases. Nature Immunol. 2019;20:1594–1602

[101] Wang T, Liu Y, Zou, J-F, Chen Z-S. Interleukin-17 induces human alveolar epithelial to mesenchymal cell transition via TGF-beta-1 mediated Smad2/3 and ERK-1/2 activation. PLOS ONE2017;12 (9): e0183972.doi.10.137/journal. pone.0183972

[102] Weng C-H, Li Yi-J, Wu H-H, Liu S-H, Hsu H-H, Chen Y-C, et al.Interleikin-17A induces renal fibrosis through ERK and Smad signaling pathways. Biomed. Pharmacother. 2020;123:109741. doi.10.1016/j.biopha.2019.109741

[103] Fuqua CF, Akomeah R, Price JO, Adunyah, SE. Evidence for the involvement of Erk1/2 in IL-21-induced cytokine production in leukemia cells and human monocytes. Cytokine. 2008; 44:101–107

[104] Phizicky EM. Protein-protein interactions: methods of detection and analysis. Microbil.Rev.1995;59:94–123

[105] Golemis E. Protein-protein interactions: a molecular cloning manual. Cold Spring Harbor (NY). 2002;Cold Spring Harbor Laboratory Press. p ix,682

[106] de Totero TD. Meazza R, Simona Zupo S, Cutrona G, Matis S, Colombo M, et al. Imterleukin-17 21 receptor (IL-21R) is up-regulated by CD40 triggering and mediates proapoptotic signals in chronic lymphocytic leukemia B cells. Blood. 2006;107:3708–3715

[107] Dignam JD, Lebovitz RM, Roeder RG. Accurate transcription initiation by RNA polymerase II in a soluble extract from isolated mammalian nuclei. Nucleic Acids Res. 1983;11:1475–1489

[108] Lafarge S, Hamzeh-Cognasse H, Charvarin P, Genin C, Garraaud O, Cognasse F. Flow cytometry technique to study intracellular signals NF-kappaB and STAT3 in peripheral blood mononuclear cells. BMC Molecular Biology. 2007;8(64)1–9

[109] Jiang H, Yu j, Guo H, Song H, Chen S. Upregulation of survivin by leptin/STAT3 signaling in MCF-7 cells. Biochem. Biophys. Res. Commun. 2008; 368(1):1–5

[110] Berridge MV, Herst PM, Tan AS. Tetrazolium dyes as tools in cell biology: New insights into their cellular reduction. Biotechnology Annual Rev. 2005:127–150

[111] Namura S, Zhu J, Fink K, Endres M, Srinivasan A, Tomaselli KJ, et al. Activation and cleavage of

Caspase-3 in apoptosis induced by experimental cerebral ischemia. Journal. Neuroscience. 1998;18(10):3659–3668

[112] Hague A, Manning AM, Hanlon KA, Huschtscha LI, Dart D, Paraskeva C. Sodium butyrate induces apoptosis in human colonic tumor cell lines in a P53-independent pathway: implications for the possible role in dietary fiber in the prevention of large-bowel cancer. Int. J. Cancer. 1993;55(3): 498–505

[113] McCubrey WA, May WS, Duronio V, Mufson A. Serine/Threonine phosphorylation in cytokine signal transduction. Leukemia. 2000;14(1): 9–21

[114] Condorelli F, Salomoni P, Cotteret S, Cesi, V, Srinivasa M, et al. Caspase enhances the apoptosis-inducing effects of BAD. Mol. Cell Biol. 2001;21(9):3025–3036

[115] Kanda N, Seno H, Konda Y, Murusawa H, Kanai M, et al. STAT3 is constitutively activated and supports cell survival in association with surviving expression in gastric cancer cells. Oncogene. 2004;23:4921–4929

[116] La Fortezza M, Schenk M, Cosolo A, Kolybaba A, Grass I, Classen A-K. JAK'STAT signaling pathway mediates cell survival in response to tissue stress. Development. 2016;143:2907–2919

[117] Nave BT, Ouvens M, Withers DJ, Alessi DR, Shepherd PR. Mammalian target of rapamycin is a direct target for protein kinase B: Identification of a convergence point for opposing effects of insulin and amino-acid deficiency on protein translation. Biochem J. 1999;344 (Pt.2):427–431

[118] de Rham C, Ferrari-Lacraz S, Jendy S, et al. The proinflammatory cytokines IL-2. IL-15 and IL-21 modulate the repertoire of mature human natural killer cell receptors. Arthritis Res. Ther. 2007;9:R125.doi.org/10.1186/ ar2336

[119] Liao W, Lin J-X, Leonard W. IL-2 Family of Cytokines: New insights into the complex roles of IL-2 as a broad regulator of helper T cell differentiation. Curr. Opin. Immunol. 2011;23(5): 598–604

[120] Ihle JN. Immunological regulation of hematopoietic stem cell function by IL-3 and its role inkeukemogenesis.IL-3 The Panspecific Hematopoietin. 1988; pages 127–161

[121] Sieff CA, Niemeyer CM, Nathan DG, Ekern SC, Yang YC, Wong G, Clark SC. Stimulation of human hematopoietic colony formation by recombinant gibbon multi-stimulating factor or Interleukin-3.J. Clin. Invest. 1987;80(3):818–829

[122] Couper KN, Blount DG, Riley EM. IL-10: The master regulator of immunity to infection.J.Immunol. 2008; 180(9):5771–5777

[123] Li X, Lu P, Li B, Zhang W, Yang R, Chu Y, Luo K. Interleukin-2 and Interleukin-10 function synergistically to promote CD+T cell cytotoxicity, which is suppressed by regulatory T cells in breast cancer. Int. J. of Biochem. & Cell Biol. 2017;87:1–7

[124] Inoue S, Unsinger J, Davis, CG, Muenzer JT, et al. IL-15 prevents apoptosis, reverses innate and adaptive immune dysfunction, and improves survival in sepsis. J. Immunol. 2010;18 (3):1401–1409

[125] Waldmann TA. The shared and contrasting roles of Interleukin-2 (IL-2) and Inteleukin-15 (IL-15) in the life and death of normal and neoplastic lymphocytes: implications for cancer therapy. Cancer Immunol. Res. 2015;3 (3):219–227

[126] Brunet A, Bonni A, Zigmond MJ, et al. Akt promotes cell survival by phosphorylating and inhibiting a Forkhead Transcription Factor. Cell. 1999;96:857–868

[127] Trizzino M, Zucco A, Deliard S, Wang F, et al. EGR-1 is a gatekeeper of inflammatory enhancers in human macrophages. Science Advances. 2021;7 (3):eaaz8836 doi.10.1126/sciadv, aaz8836

[128] Yang H, Wang L, Zhao J, et al. TGF-beta-activated smad3/4 complex transcriptionally upregulates N-cadherin expression in non-small cell lung cancer. Lung Cancer. 2015;87(3): 249–257

[129] Williams JCM. Evidence of the involvement of specific Src kinases and phospholipase D in IL-17A signaling mechanisms in U937 leukemia cells [thesis]. Biochemistry and Cancer Biology Department: Mehary Medical College, Nashville, TN; 2004

[130] Davis WL, Jr. An in vitro examination of the effects of Interleukin-17 in proliferation and differentiation of U937 leukemia cells [thesis]. Biochemistry and Cancer Biology Department: Mehary Medical College, Nashville, TN; 2004

[131] Wu D, Yang XO. TH17 responses in cytokine storm of COVID-19: An emerging target of Jak2 inhibitor Fedratinib. J. MicroBiol. Immunol. & Infect. 2020;53:368–370

Chapter 7

The Role of Interleukins after Spinal Cord Injury

Daniel J. Hellenbrand, Rylie M. Roddick, Sophia M. Mauney, Ryan T. Elder, Carolyn N. Morehouse and Amgad S. Hanna

Abstract

In skin wound healing the injured tissue goes through a normal progression, inflammation subsides and remodeling occurs. However after spinal cord injury inflammation persists and there is less progression into a regenerative/rebuilding phase. This inflammatory process after spinal cord injury is orchestrated by many cell types and numerous cytokines. Although there are several positive effects of inflammation after spinal cord injury, such as the removal of debris, the substantial upregulation of immune cells has been shown to contribute to neural degeneration. Several chemokines and cytokines including many interleukins are involved in guiding these immune cells to the lesion. While there are many inflammatory cytokines acting on these immune cells after SCI, there are also several anti-inflammatory interleukins that have shown beneficial effects in reducing inflammation. After SCI in a rat model, interleukin-10 and interleukin-19 have been shown to downregulate the synthesis of pro-inflammatory species including interleukin-1β and tumor necrosis factor-α, which resulted in a significant improvement in rat hind limb function. Also, interleukin-4 and interleukin-13 are related anti-inflammatory cytokines that regulate many aspects of inflammation and have also been shown to induce alternative macrophage activation. The differing and complex roles interleukins play, highlight their importance on the inflammation that persists after spinal cord injury. Here we review both the positive effects and negative effects that interleukins have during the multifaceted inflammation process following spinal cord injury.

Keywords: interleukins, spinal cord injury, inflammation, macrophages, microglia

1. Introduction

Spinal Cord Injury (SCI) is a devastating trauma and according to the National Spinal Cord Injury Statistical Center (NSCISC) there are approximately 294,000 people living with SCI in the United States [1]. After spinal SCI there is immediate cell death caused directly from the insult followed by a cascade of inflammation that leads to additional cell death and a much larger scar formation that impedes axonal regeneration [2, 3]. Although there are several positive effects of inflammation after SCI, the extensive infiltration of immune cells is a principal contributor to neural degeneration [4, 5]. These immune cells are guided to the lesion site from the periphery via many signaling cues including several interleukins (ILs) released by microglia, astrocytes, and peripheral macrophages within the lesion [5, 6].

IntechOpen

Throughout the first hours after injury, polymorphonuclear leukocytes are the predominant infiltrating cells and over-activation of these cells causes tissue destruction through the release of significant amounts of neurotoxins including reactive oxygen species (ROS), reactive nitrogen species (RNS), chemokines, and enzymes [5, 7, 8]. Microglia, the resident macrophages, are also activated and migrate to the site of injury, proliferate, and transform from the ramified phenotype to amoeboid phagocytic cells [9]. These activated microglia and peripheral macrophages make up the majority of inflammatory cells present at the site of the lesion. Although in normal wound healing macrophages sequentially change and reduce inflammation, after SCI macrophages persist in an inflammatory state for prolonged periods resulting in progressive tissue degeneration [10, 11]. However these microglia/macrophages can be activated toward an anti-inflammatory phenotype and ILs are important signaling cues in the extracellular environment that help dictate this contrasting phenotype. The goal of this chapter is to examine the role ILs have on the dynamic inflammatory process that occurs after SCI.

2. Interleukins involved in inflammation after SCI

There are numerous known ILs and several of these ILs are shown to be involved in inflammation after SCI. Throughout this chapter we will discuss the ILs that have been investigated after SCI and whether their role is predominately inflammatory or anti-inflammatory. It is important to note, the role an IL plays after injury is not as simple as just inflammatory or anti-inflammatory. For many of the ILs there are multiple factors that determine whether they will have a beneficial role or a detrimental role, including the extent of initial injury, concentration of IL, other associated molecules in the injury, and the response of immune cells and glial cells [12, 13].

Although there is a broad spectrum of signaling molecules including cytokines, chemokines, and other reactive species after SCI, this chapter will just focus on ILs and only the ILs that are known to play a role in inflammation after SCI [14]. These ILs will be discussed in terms of the cell types that produce them, receptors they bind, cell types they target, timeline of upregulation, and ultimately their effect on inflammation after SCI.

2.1 Interleukin-1 family cytokines

IL-1α, IL-1β, and IL-33 are members of the IL-1 family that have been studied after SCI and all are predominantly inflammatory [15]. IL-1 is released via activated macrophages and microglia largely in response to disease, infection, or inflammatory events. IL-1 has two structurally and biologically similar isoforms, IL-1α and IL-1β [13]. These two isoforms share roughly 30% amino acid sequence homology and although they perform similar biological functions, IL-1β plays a more substantial role post-SCI [16, 17]. IL-1β has been shown to contribute to the exaggerated neuroinflammation following SCI that leads to secondary neural degeneration and cell death [16]. IL-1 signaling following SCI is diverse and complex, resulting in a recruitment of neurotoxins or immune system molecules that contribute to the inflammatory response [13].

The primary receptor for signaling of both IL-1 isoforms is the type-I interleukin-1 receptor (IL-1RI). IL-1 signaling is further regulated by a decoy receptor (IL-1RII) and a receptor antagonist (IL-1ra) [13]. The expression of these receptors mediates the inflammatory response to SCI and their mechanisms have been widely studied following SCI. After SCI in rats, IL-1R1 expression is elevated as early as 4 hours, peaks at 8 hours to 1-day and remains elevated for 7 days post-injury [13]. Another study tested the role of IL-1ra as a regulatory molecule following SCI and observed that increased expression of IL-1ra suppressed IL-1β levels and increased locomotor function following SCI, suggesting that IL-1β and IL-1RI play critical roles in secondary tissue damage and impaired functional recovery post-SCI [16]. Likewise, administration of IL-1β suppressed the expression of IL-1ra following SCI indicating the regulatory nature of IL-1β interactions with its receptor antagonist [16]. Similarly, a study using IL-1 knockout mice observed a significantly smaller lesion area and improved locomotor function after SCI in comparison to wild-type mice [18].

The signaling cascade of the IL-1β/IL-1RI pathway is complex and yet to be completely understood, however it is widely understood that it stimulates the production of toxic intermediates that cause neural degeneration and cell death [13]. These toxic inflammatory mediators include prostaglandins, cyclooxygenase 2, and phospholipase A2 [13]. However, there are studies showing benefits of IL-1, where IL-1β null mice failed to remyelinate as rapidly as wild-type mice [19]. These different roles IL-1 plays are likely due to several factors including extent of injury as well as IL-1 concentration and timing of upregulation, but at present are not well understood.

Another member of the IL-1 family, IL-33, predominantly induces type-2 immune responses against allergens and infectious diseases [20]. IL-33 is upregulated in response to SCI and tends to localize in spinal cord astrocytes to reduce T cell infiltration and overexaggerated inflammation that leads to neuronal cell death [21]. IL-33 is classified as an alarm signal (alarmin) and is released by epithelial cells upon signals of cell or tissue death, but the exact *in vivo* mechanism of release is not fully understood [22]. After its release, IL-33 binds to ST2 receptors (IL-1RL1) that are present on multiple immune cells as an alert signal for immunologic and neurologic damage or inflammation [22].

One study that treated SCI injury in mice with administration of recombinant IL-33 indicated an attenuation of spinal cord encephalomyelitis progression and a significant decrease in neural tissue death, decrease in demyelination, and an overexaggerated astrocyte infiltration at the lesion site of the contused spinal cord [21]. These results yielded a significant increase in functional recovery and a dramatic decrease of the expression of TNF-α in the spinal cord for as long as 42 days post-SCI. In addition to suppression of pro-inflammatory cytokine release, IL-33 administration promoted the activation of anti-inflammatory M2 macrophage/microglia [21].

2.2 Interleukin-2 family cytokines

Cytokines from the IL-2 family, IL-2, IL-4, IL-7, IL-15, and IL-21, all share a common receptor subunit (gammac), which plays a major role in promoting and maintaining T lymphocyte populations [23]. IL-2 is a pro-inflammatory cytokine made up of four α helixes and is produced mainly by $CD4^+$ cells when activated. At an mRNA level, signals from T-cell receptor (TCR) and CD28 closely regulate the production of IL-2 [24]. After synthesis, IL-2 binds to a receptor complex, which consists of three subunits, IL-2Rα, IL-2Rβ, and the common γ-chain [24]. All three subunits are needed to achieve high affinity binding. These receptors are located on regulatory T cells and antigen-activated T lymphocytes [25]. To produce an IL-2-dependent response, IL-2 must be produced and IL-2R must be expressed within the same microenvironments [25].

IL-2 and its receptor, IL-2R, are crucial to maintaining the balance of the timing and adequacy of an immune response [26]. The primary role of IL-2 is to perpetuate the proper response of memory T-cells to invading pathogens [27]. In addition, IL-2 is vital to the survival, as well as death, of lymphocytes, which has an effect on the development of the immune system. By properly maintaining the life of regulatory T cells (T reg) and activation-induced cell death, IL-2 is able to eliminate self-reactive T cells as a preventative measure against autoimmune diseases [27]. After SCI in a rat, IL-2 levels were significantly lower than intact controls from 3 days to 2 weeks post-SCI [14]. In addition, the interaction of IL-2 with its receptor after SCI contributes to the proliferation of T-helpers, which also have an effect on the proliferation of cytotoxic T cells, natural killer cells, lymphokine-activated killers, B cells, and macrophages [14].

IL-4 and IL-13 are related anti-inflammatory cytokines that regulate many aspects of inflammation and have also been shown to induce alternative macrophage activation (**Figure 1**) [28]. IL-4 is a cytokine that is involved in regulating immunity, and is secreted by Th2 cells, eosinophils, basophils, and mast cells [29]. IL-4 is also involved in allergic inflammation by utilizing Th2 lymphocytes, differentiated from Th cells, which can then be used in the production of effector cytokines [30]. IL-4 binds to its receptor IL-4Rα, and will dimerize with either γc (the common cytokine-receptor γ-chain) and produce the type-1 signaling complex, or with IL-13Rα1 and produce the type-2 signaling complex (**Figure 1**) [29, 31]. Although IL-4 has a major impact on immunity, it also affects cognition based on T-cells mediated by IL-4. When administered within a short period post injury, IL-4 exhibits anti-inflammatory effects; however, it can exert a pro-inflammatory response when macrophages possessing IL-4 are undergoing pro-inflammatory stimulation [29].

Lima et al. (2017) performed a study to understand the effect of the acute and sub-acute treatment using IL-4 on various populations of neural cells and on functional recovery *in vivo*. In the injured spinal cord, treatment using a systemic delivery of IL-4 (0.35 μg/kg) for 7 days, led to an upregulation of the anti-inflammatory

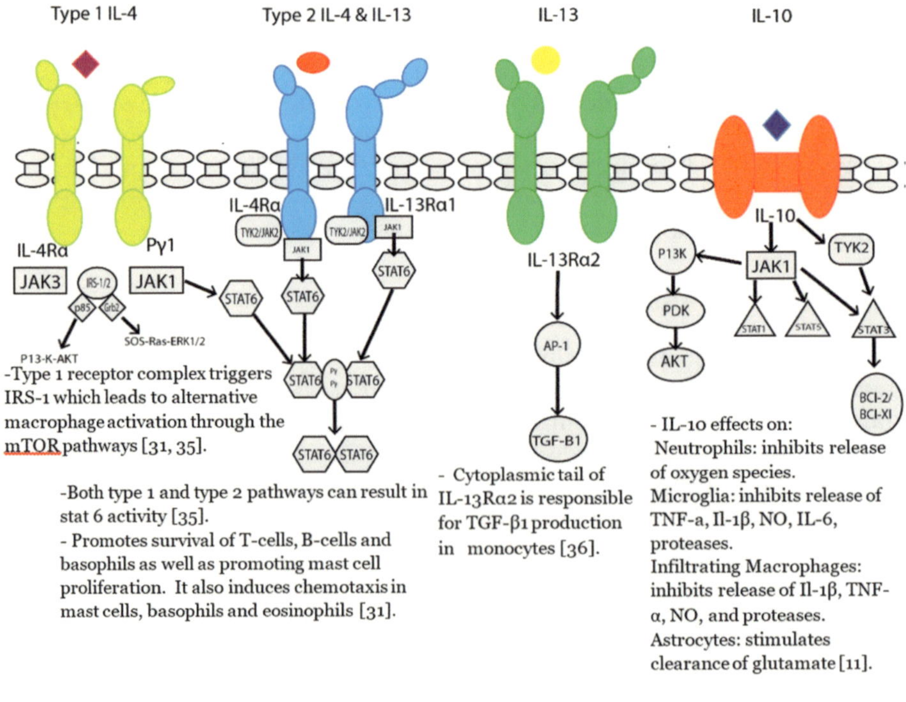

Figure 1.
IL-4, IL-13, and IL-10 pathways and effects.

IL-10 and a reduction in area of macrophage/microglia expressing inflammation markers CD11b and inducible nitric oxide synthase (iNOS) [32]. After systemic IL-4 treatment, they also observed an increase in the number of O4-positive cells (a marker for both type I and type II oligodendrocytes) and neuronal markers βIII-tubulin and NeuN, suggesting that IL-4 has a role in neuroprotection. This overall reduction in inflammation resulted in improved hind limb function in rats after SCI. Although they observed several positive effects, systemic IL-4 did not have an effect on the number of astrocytes or lesion size [32]. In another study a delayed intraspinal injection of IL-4 (100 ng of recombinant IL-4, 48 hours after injury) was given after a spinal cord contusion in mice [33]. The intraspinal injection of IL-4 resulted in an increase in microglia/macrophages expressing antigens characteristic of an anti-inflammatory M2 phenotype, reduced tissue damage, and improved hind limb function in mice after SCI [33]. These studies suggest that therapies using IL-4 could be a valuable treatment for improving function after SCI.

IL-13 is produced by different T cell subsets, dendritic cells, and activated Th2 cells [34]. Although the IL-13α2 receptor was originally thought to be a "decoy" receptor that serves as a neutralizer (**Figure 1**) [35], Fichtner-Feigl et al. showed a role for IL-13Rα2-mediated signaling that required the cytoplasmic tail of IL-13Rα2 in the production of transforming growth factor beta (TGF-β), an anti-inflammatory shown to down-regulate inflammatory cytokines, providing evidence for IL -13Ra2-mediated signaling (**Figure 1**) [31, 36]. Furthermore, after SCI it was shown that transplanted mesenchymal stem cells continuously expressing IL-13 improved functional recovery and decreased lesion size. In addition, IL-13 increased the amount of ARG-1-expressing macrophages [37].

IL-7 is a homeostatic cytokine that plays a key role in the survival of multiple immune cells and acts on lymphocytes [38]. The IL-7 receptor complex is composed of two chains, IL-7Rα and γc (the common cytokine-receptor γ-chain), which signal downstream to the JAK/STAT5 pathway, and assists in regulating the survival and development of immune cells [38]. IL-7 is produced by stromal cells in lymphoid organs and is necessary for T-cell development and their survival in the periphery [39].

After SCI in mice, IL-7 is promptly upregulated and displays as a strong chemotactic property for macrophages [40]. An intraspinal injection of IL-7 after SCI in mice, resulted in an increase in pro-inflammatory cytokines IL-1β, IL-6, and TNF-α, and a decrease in the anti-inflammatory cytokine IL-10 [38]. The increase in IL-7 also led to an increase in apoptosis, macrophage infiltration, and a decrease in hind limb function in mice after SCI [38]. Moreover, blocking the IL-7 receptor after SCI in mice, resulted in suppression of pro-inflammatory cytokines IFN-γ and TNF-α, an increase in IL-4 and IL-13, more macrophages expressing antigens characteristic of an anti-inflammatory M2 phenotype, an increase in spared white matter, and an improvement in hind limb function [40]. During SCI, the JAK/STAT5 pathway is activated, and IL-7 post-SCI also contributes to the activation of the JAK/STAT5 pathway, which upholds a crucial role in the inflammatory response and secondary damage [38]. When the JAK/STAT5 pathway was inhibited by pimozide, the effects of IL-7 discontinued, which emphasizes the relationship between the JAK/STAT5 pathway and IL-7 function [38]. Therefore, Yuan et al. (2019) concluded that the IL-7/JAK/STAT5 axis targeted by antagonists may represent a potential therapeutic treatment for SCI [38].

Similar to IL-2, IL-15 is a pro-inflammatory cytokine that is also part of the four α helix cytokine family. The main function of IL-15 is to provide a long-term immune response to invading pathogens by contributing to the homeostasis of natural killer cells and CD8+ memory T cells that express IL-2/IL-15Rβ and γc [27].

IL-15 has three receptors, IL-15Rα, IL-2Rβ, and γc, and shares two of the receptors with IL-2 (IL-2Rβ and γc) [41]. Although IL-15 has not been well studied after SCI, it has been shown to be involved in the development of neuropathic pain from nerve injury [42]. After sciatic nerve injury, IL-15 expression was observed in the spinal cord in astrocytes and microglia, and it is also present in neurons located in the dorsal and ventral horn [42].

IL-21 is a pleiotropic cytokine expressed by many immune cells including natural killer T cells and activated CD4+ T cells [43]. Similar to other inflammatory mediators, IL-21 is upregulated after SCI [44, 45]. Fu et al. (2017) studied peripheral blood-derived mesenchymal stem cells (PBMSCs) as a therapy for SCI and their role in the lesion microenvironment by analyzing the neuroprotection, differentiation, and immunoregulation of PBMSCs that were engrafted. When IL-21 was inhibited, a decrease in the secretion of IL-23a and IL-22 occurred [44]. When investigating the potential Th17/Treg-relative mechanism of PBMSCs therapy after SCI, Fu et al. (2017) discovered that the M1 macrophage migrated to lesion site and resulted in the pro-inflammatory secretion of IL-6 and IL-21, which led to CD4 + T cells differentiating into CD4 + IL17 + Th17 cells [44]. Furthermore it has been shown that IL-17 production is stimulated by the combination of IL-21 and TGF-β [45].

2.3 β common chain cytokines IL-3, and IL-5

The β common cytokine family, including IL-3 and IL-5, is defined by a shared receptor structure, comprising of a specific α chain and a common β chain that is essential for cytokine-specific receptor signaling [46]. IL-3 is a cytokine that is produced by activated T cell lymphocytes, which then induces the production of various hematopoietic cell types that are crucial to the immune response [47]. IL-5 is cytokine that is produced by hematopoietic and non-hematopoietic cells, including granulocytes, T cells, and natural helper cells [48]. IL-5 is also a mediator for eosinophilic inflammation by providing stimulation, differentiation, recruitment and activity of eosinophils. Due to their roles with eosinophils, IL-3 and IL-5 have been primarily studied in asthma, and their roles after SCI are not clear. However, both of them co-express in TH2 cells, which is a subset of CD4+ cells. These TH2 cells are characterized by the production of IL-4, IL-5, IL-10 and IL-13 and thus, may be beneficial in exerting anti-inflammatory effects after SCI [49, 50].

2.4 IL-6 and IL-11

IL-6 and IL-11 are grouped into one cytokine family because the receptor complex of each cytokine contains two of the signaling receptor subunit gp130 (**Figure 2**) [51]. For both IL-6 and IL-11 there are membrane bound receptors as well as soluble receptors and after ligand binding to either the membrane bound receptor or the soluble receptor, they form a complex with two gp130 receptors leading to Jak/STAT pathway signaling (**Figure 2**) [52, 53].

IL-6 is predominantly an inflammatory cytokine. After SCI in a mouse model, Pineau et al. observed IL-6 mRNA expression in astrocytes, microglia/ macrophages, and neurons, starting at 3 hours post-injury, peaking at 12 hours and continuing for 4 days post-injury [54]. Similarly after SCI in humans IL-6 is strongly upregulated. IL-6 levels in cerebrospinal fluid of SCI patients changed from undetectable (<4 pg./ml) in non-injured controls to an average of almost 30,000 pg./mL in the subset of patients with complete SCI [55]. Furthermore, the cerebrospinal levels of IL-6 correlated with the extent of spinal cord damage in humans, which demonstrates the importance of IL-6 after SCI.

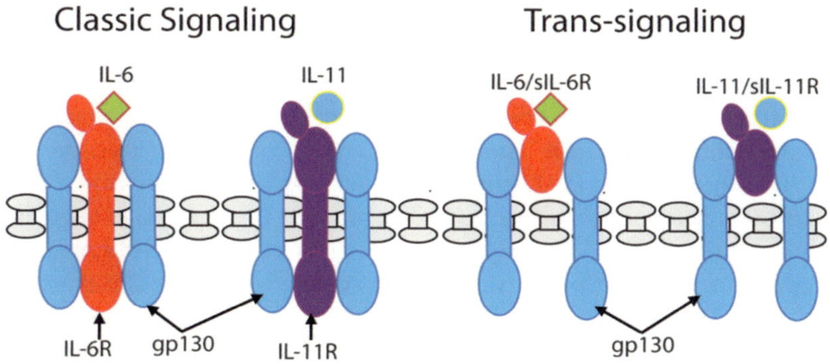

Figure 2.
Classic and trans-signaling. Cells that express the IL-6R or IL-11R will undergo classic signaling when IL-6 or IL-11 bind to the corresponding receptor, inducing gp130 dimerization and initiating intracellular signal transduction. Trans-signaling occurs when the ligand and soluble receptor complex (IL-6/sIL-6R or IL-11/sIL-11R) associate with gp130 inducing gp130 dimerization and initiating intracellular signal transduction.

IL-6 leads to recruitment of immune cells. Delivery of IL-6/sIL-6 receptor fusion protein to injury sites induced a sixfold increase in neutrophils and a twofold increase of macrophages/microglial [56]. Mice treated with an antibody against IL-6 receptor showed a reduction in neutrophil and monocyte/macrophage invasion [57, 58]. It was also shown that blocking IL-6 signaling after SCI reduces the damaging inflammatory activity by promoting the formation of alternatively activated M2 macrophages [59]. Taken together these data suggest IL-6 signaling is an activator of inflammation and a strong recruiter of immune cells after SCI.

After SCI astrocytes proliferate and migrate to the injury leading to a dense astroglial scar surrounding the lesion. It has been shown *in vitro* that IL-6 signaling acts on neural stem cells to induce their differentiation into astrocytes [60]. This was supported by several *in vivo* studies including, IL-6 knockout mice that showed suppression of astrogliosis following SCI [61], mice with an excessive expression of IL-6 and IL-6R showed abundant astrogliosis suggesting that astrocytes were selectively affected in these mice [62], and the development of astrogliosis was inhibited in mice given an IL-6 receptor blocker after SCI [58].

Several studies have shown that blocking IL-6 signaling improves functional recovery after SCI [57–59]. It has also been shown that delivery of IL-6/IL-6 receptor resulted in a four fold decrease in axon growth [56]. However there are studies showing that IL-6 is neuroprotective and aids in axonal regeneration [63, 64]. The differences in IL-6 effect may depend on the level of expression and timeline of IL-6 upregulation. The studies using an IL-6 blocker were performed in the sub-acute timeframe after SCI. In the sub-acute SCI, any neurotrophic effects of IL-6 appear to be overwhelmed by its proinflammatory features. Taken together, the aforementioned data demonstrates the importance of IL-6 after SCI. IL-6 upregulates inflammatory cytokines, recruits immune cells, effects macrophage phenotype, effects astrocyte activation, effects axonal regeneration, and effects functional recovery.

IL-11 has been shown to be primarily anti-inflammatory. Recombinant IL-11 administered to activated macrophages inhibited the production of pro-inflammatory cytokines TNF-α, IL-1β, IL-12, and nitric oxide production [65–67]. Furthermore, IL-11 has been shown to play an anti-inflammatory role in the airways for asthma [68], play a role in decreasing mucosal damage in inflammatory bowel disease [69], and importantly IL-11 has a neuroprotective role in multiple sclerosis [70]. Due to these anti-inflammatory roles, Cho et al. analyzed the role of IL-11 after SCI using IL-11Rα knockout mice [71]. In wild type mice, they observed a significant upregulation in IL-11 with a peak gene expression 24 hours after injury and a significant upregulation of IL-11Rα at 3 and 7 days after SCI. Somewhat surprisingly, they did not observe significant differences in functional recovery or histopathology in IL-11Rα knockout mice as compared to wild type mice after SCI. The authors speculate that since "the peak in IL-11Rα expression is on the order of days after SCI, suggests that IL-11 signaling may not play as significant a role in the acute inflammatory response after injury, but more in the long-term sequelae such as oligodendrocyte survival". Maheshwari et al. used a cuprizone induced mouse model of demyelination in the central nervous system to analyze the effects of over-expression of IL-11 on demyelination/remyelination [72]. Overexpression of IL-11 was able to limit cuprizone-induced demyelination by reducing oligodendrocyte cell death, decrease microglial activation, and enhance spontaneous remyelination. Maheshwari's results further suggest that IL-11 likely plays a role in the long-term remyelination efforts after SCI and is not as involved in the sub-acute stage [72].

2.5 Interleukin-8 and interleukin-16

IL-8, also known as neutrophil chemotactic factor or CXCL8, primarily induces chemotaxis in neutrophils and granulocytes. IL-8 is a member of the chemokine family that acts on CXCR1 and CXCR2 receptors (il8ra and il8rb, respectively), which have been primarily studied on polymorphonuclear leukocytes. However many other cell types express these receptors including neurons [73]. Several studies have shown that IL-8 can be released by a wide variety of cells including monocytes endothelial cells, T lymphocytes, and macrophages [73]. After SCI in rat, GRO, the rat analogue of human IL-8, is strongly upregulated for at least 14 days and the upregulation of GRO strongly correlates with the extent of injury [14, 74, 75]. Furthermore IL-8 is upregulated in the cerebrospinal fluid of dogs and humans after SCI, and for humans the IL-8 levels are also shown to correlate with the extent of damage [55, 76, 77]. Although IL-8 clearly plays a role in neutrophil infiltration and overall inflammation after SCI, as shown by its significant upregulation, it has not been extensively studied after SCI.

IL-16 is a proinflammatory cytokine that is produced by mast and leukemic cells, fibroblasts, endothelial cells, granulocytes, dendritic cells, CD4+ and CD8+ T lymphocytes, monocytes, and microglial cells. IL-16 plays a role in the release of other proinflammatory cytokines (IL-1β, IL-6, IL-15, and TNFα), the increase of intracellular Ca^{++} or inositol-(1,4,5)-triphosphatase, and the translocation of protein kinase C [78]. These processes occur after IL-16 binds to the signal-transducing CD4 receptor molecule [79]. Moreover, IL-16 promotes lymphocyte migration and modulates apoptosis [80].

Following spinal cord injury, IL-16 plays a role in recruiting and activating inflammatory cells. Microglia that produce IL-16 migrate to the lesion site and other areas of significant neuronal damage [78]. Following neuroinflammation, it is suggested that IL-16 microglia are one of the first cells to respond [80]. In addition, macrophages with IL-16 remained present at the injury site for up to thirty days post injury, indicating long-term IL-16 function [78]. One study found that expression of IL-16 in microglia and macrophages is induced by the IL-12 p40 homodimer through IL-12Rβ1, but not IL-12 p70 [80]. Overall, the ability of IL-16 to quickly recruit microglia/macrophages to the lesion site following SCI results in increased neuronal damage and microvessel clustering [78].

2.6 Interleukin-10 family cytokines

Members of the IL-10 family of cytokines that have been studied after SCI include, IL-10, IL-19, IL-20, and IL-22 [81]. IL-10 is an anti-inflammatory cytokine that is produced by monocytes, B cells, dendritic cells, natural killer cells, and T cells [82]. In leukocytes, IL-10 acts on both innate and adaptive immune cells with a wide range of immunomodulatory activities that suppress proliferation, cytokine secretion, and costimulatory molecule expression of proinflammatory immune cells. The IL-10 receptor consists of heterotetramer complex made of two IL-10R1 molecules, encoded by the IL10ra gene, and two IL-10R2 molecules, encoded by the IL10rb gene (**Figure 1**) [83]. IL-10 downregulates several pro-inflammatory cytokines and inflammatory species [11]. In addition, IL-10 can affect T cell and natural killer cell function indirectly and directly through connection with monocytes and macrophages. The overall impact of IL-10 is determined by the timing and site of its production, which are both affected by which cells are making IL-10. Since IL-10 production by one cell type affects the ability of other cells to make IL-10, IL-10-producing cells show potential to regulate each other [82].

Following SCI, IL-10 downregulates pro-inflammatory molecules IL-1β, IL-2, IL-6, TNF-α, IFN-γ, matrix metalloproteinase-9, nitric oxide synthase, myeloperoxidase, and reactive oxygen species. IL-10 also provides trophic support to neurons through downregulation of pro-apoptotic factors cytochrome c, caspase 3, and Bax, as well as upregulation of anti-apoptotic factors B cell lymphoma 2 (Bcl-2) and Bcl-2-associated X, B-cell lymphoma-extra large (Bcl-xl) (**Figure 1**) [11]. There have been several studies performed to test IL-10's therapeutic value as a treatment for SCI. Although these studies used a variety of different systemic and local methods to administer IL-10 after SCI, the majority of results showed strong positive effects from the IL-10. These positive effects after SCI include a reduction in pro-inflammatory molecules, macrophages expressing more antigens characteristic of an anti-inflammatory M2 phenotype, reduced lesion size, and an improvement in hind limb function [11, 84].

IL-19 is produced by monocytes and microglia, and binds to the IL-20 receptor complex, which consists of IL-20R1 and IL-20R2 chains [85]. Activated microglia upregulate IL-19 and express the IL-20 receptor complex [86]. It has also been shown that ablation of IL-19 in activated microglia increased the production of pro-inflammatory cytokines IL-6 and TNF-α, which demonstrates that IL-19 is predominately an anti-inflammatory cytokine in the central nervous system [86].

After SCI in mice, IL-19, IL-20R1 and IL-20R2 are upregulated [87]. In a series of four different experiments, mice with spinal cord injuries were treated with IL-19 [87]. As a result, Th2 cytokine synthesis was promoted, which polarized spinal microglial cells to an M2 phenotype. This helped resolve the inflammation, preserving myelin, neurons, and neuronal function. Overall, IL-19 attenuated macrophage accumulation, reduced protein levels of TNF-α and CCl2, promoted Th2 response and M2 macrophage activation, promoted angiogenesis by upregulating VEGF, upregulated HO-1 expression, and decreased oxidative stress in the injured region [87].

IL-20 is a proinflammatory cytokine that is predominately produced by monocytes and skin keratinocytes. IL-20 signals through both the IL-20R1/IL-20R2 heterodimer complex and the receptor complex composed of IL-22R1 and IL-20R2 [85]. Following spinal cord injury, IL-20 and its receptors are expressed in neurons, astrocytes, oligodendrocytes, and microglia in large amounts. IL-20 upregulates glial fibrillary acidic protein (GFAP), TGF-β1, TNF-α, MCP-1, and IL-6 expression, which stimulates astrocyte reactivation and migration [88]. As a result, glial scar border formation is enhanced. Moreover, IL-20 inhibits neuron outgrowth through upregulation of Sema3A/NRP-1 in PC-12 cells [88]. The overall result is irreversible neuronal loss and glial scar formation post-SCI. *In vivo*, anti-IL-20 mAb reduces the IL-20 inflammatory response, which improves motor and sensory functions, spinal cord tissue preservation, and reduces glial scar formation [88].

2.7 Interleukin-12 family cytokines

The IL-12 family is comprised of 4 members, IL-12, IL-23, IL-27 and IL-35 and each member is composed of α-subunit with a helical structure similar to type 1 cytokines and a β-subunit structurally related to the extracellular regions of Type 1 cytokine receptors [89]. However from this family of 4 cytokines, only the pro-inflammatory cytokine IL-12 has been assessed after SCI. IL-12 is produced by dendritic cells, macrophages, monocytes, neutrophils, microglia cells, and B cells [90]. The IL-12 receptor is made up of IL-12Rβ1 and IL-12Rβ2 chains [91]. IL-12 is a heterodimeric molecule, p70, formed from p40 and p35 chains. IL-12p70 is considered to be the biologically active cytokine that expresses nitric oxide synthase and TNF-α in microglia and macrophages. In T cells, p70 interacts with both IL-12Rβ1 and IL-12Rβ2. However, p70 treatment results in IL-16 mRNA inhibition due to inability to induce IL-16 promoter [80].

Yaguchi et al., administered IL-12 after SCI in mice and observed an increase in the number of activated macrophages and dendritic cells surrounding the lesion site and an increase in the expression of brain-derived neurotrophic factor adjacent to the injury. After IL-12 treatment, immunohistochemical analyses revealed that *de novo* neurogenesis and remyelination occurred. The mice treated with IL-12 also had a significant improvement in hind limb function [92].

2.8 Interleukin-17 family cytokines

IL-17 cytokines play important roles in both innate and adaptive immunity. IL-17A to IL-17F are highly conserved at the C terminus, and contain five spatially conserved cysteine residues that mediate dimerization [93, 94]. IL-17A and IL-17E have been identified and studied for the roles they play after SCI.

IL-17A is an important cytokine in regard to protective mechanisms against infectious diseases and inflammatory pathology within the immune system [95]. IL-17A is secreted by a multitude of cells including T cells, dendritic cells, and macrophages among others and binds to the A and C subunits of the IL-17 receptor to initiate signaling [95]. After SCI in rats, IL-17A is upregulated as early as 1 hour after injury, peaks at 24 hours, and remains above normal levels for at least 72 hours after injury [45]. This upregulation of IL-17 does appear to play a degenerative role on SCI recovery after a study was conducted using IL-17 knockout mice. IL-17 knockout mice showed increased locomotor function and decreased lesion size after SCI, which suggests that IL-17 expression regulates secondary degeneration of the neural tissue at the lesion site [96]. Recruitment of immune cells such as B cells, neutrophils, and dendritic cells were downregulated at 6 weeks following SCI [96].

Interleukin-25, also known as Interleukin-17E, is in the IL-17 family and binds to the heterodimer complex of IL-17A and IL-17-B receptor subunits. IL-25 has primarily been understood as a systemic type-2 inflammatory mediator that triggers significant helper T-cell expression and proinflammatory cytokine suppression, however its response following spinal cord injury is largely unknown. IL-25 is primarily derived from epithelial cells and macrophages in response to infection or inflammation and contributes to type-2 helper T cell (Th-2) activation [97]. Th2 cells are responsible for the release of anti-inflammatory cytokines IL-4, IL-5, and IL-13 which play a role in neural protection and regeneration against inflammation and neurotoxins [97].

The trafficking mechanism and inflammatory response of IL-25 post-SCI remains relatively unclear, but local injection of IL-25 into the lesion site post-SCI yields interesting and contradictory results. The local administration of IL-25 following spinal cord injury in 10-week old mice results in decreased locomotor function, an increase in lesion size, and neuronal demyelination which contradicts the systemic immune response upon an IL-25 presence [97]. Interestingly, the systemic administration of IL-25 show ineffective results in regard to improved functional mobility following spinal cord injury. Microglia and astrocytes survival are also unaffected upon injection of IL-25 suggesting that IL-25 indirectly activates inflammatory molecules associated with these immune events [97]. These results raise questions about the precise role of IL-25 after SCI and possible therapeutic interventions using IL-25.

3. Conclusions

Although significant progress has been made in terms of spinal stabilization and medical care of patients after SCI, there has not been much progress made in terms of treatments for SCI to retain or regain the function that is lost. In order to design treatments for SCI, a better understanding of the inflammation process is crucial. As outlined in this chapter ILs are an intricate player in inflammation after SCI. For some of these ILs, there timeline of involvement and roles they play in inflammation has been defined. However, there is still much more research that needs to be completed to understand the roles many of these ILs play. Along with understanding the current ILs, there will assuredly be more signaling cues discovered that are involved after SCI.

Could inflammation be modulated to retain or regain a significant amount of function after SCI? This is a fundamental question that needs to be addressed. As highlighted in rodent models, such as what is observed in IL-17 knockout mice or treatments with anti-inflammatory cytokines, modulating inflammation is a promising approach for treating SCI. However it is important to realize that all variables including age, sex, level of injury, and force to cause the trauma, are controlled in these rodent models, and thus treating human SCI will be more challenging. These facts highlight the essential need to conduct more research on inflammation after SCI.

Author details

Daniel J. Hellenbrand, Rylie M. Roddick, Sophia M. Mauney, Ryan T. Elder, Carolyn N. Morehouse and Amgad S. Hanna*
Department of Neurological Surgery, University of Wisconsin, Madison, WI, USA

*Address all correspondence to: hanna@neurosurgery.wisc.edu

References

[1] 'National Spinal Cord Injury Statistical Center, Facts and Figures at a Glance', (2020).

[2] McKeon R. J., Schreiber R. C., Rudge J. S., and Silver J., 'Reduction of Neurite Outgrowth in a Model of Glial Scarring Following Cns Injury Is Correlated with the Expression of Inhibitory Molecules on Reactive Astrocytes', *J Neurosci,* 11 (1991), 3398-411.

[3] Fawcett J. W., and Asher R. A., 'The Glial Scar and Central Nervous System Repair', *Brain Res Bull,* 49 (1999), 377-91. DOI: 10.1016/s0361-9230(99)00072-6.

[4] Glaser J., Gonzalez R., Perreau V. M., Cotman C. W., and Keirstead H. S., 'Neutralization of the Chemokine Cxcl10 Enhances Tissue Sparing and Angiogenesis Following Spinal Cord Injury', *J Neurosci Res,* 77 (2004), 701-8. DOI: 10.1002/jnr.20204.

[5] Garcia E., Aguilar-Cevallos J., Silva-Garcia R., and Ibarra A., 'Cytokine and Growth Factor Activation in Vivo and in Vitro after Spinal Cord Injury', *Mediators Inflamm,* 2016 (2016), 9476020. DOI: 10.1155/2016/9476020.

[6] Mortazavi M. M., Verma K., Harmon O. A., Griessenauer C. J., Adeeb N., Theodore N., and Tubbs R. S., 'The Microanatomy of Spinal Cord Injury: A Review', *Clin Anat,* 28 (2015), 27-36. DOI: 10.1002/ca.22432.

[7] Stirling D. P., and Yong V. W., 'Dynamics of the Inflammatory Response after Murine Spinal Cord Injury Revealed by Flow Cytometry', *J Neurosci Res,* 86 (2008), 1944-58. DOI: 10.1002/jnr.21659.

[8] Taoka Y., Okajima K., Uchiba M., Murakami K., Kushimoto S., Johno M., Naruo M., Okabe H., and Takatsuki K., 'Role of Neutrophils in Spinal Cord Injury in the Rat', *Neuroscience,* 79 (1997), 1177-82.

[9] Kim S. U., and de Vellis J., 'Microglia in Health and Disease', *J Neurosci Res,* 81 (2005), 302-13. DOI: 10.1002/jnr.20562.

[10] Gensel J. C., and Zhang B., 'Macrophage Activation and Its Role in Repair and Pathology after Spinal Cord Injury', *Brain Res,* 1619 (2015), 1-11. DOI: 10.1016/j.brainres.2014.12.045.

[11] Thompson C. D., Zurko J. C., Hanna B. F., Hellenbrand D. J., and Hanna A., 'The Therapeutic Role of Interleukin-10 after Spinal Cord Injury', *J Neurotrauma,* 30 (2013), 1311-24. DOI: 10.1089/neu.2012.2651.

[12] Sochocka M., Diniz B. S., and Leszek J., 'Inflammatory Response in the Cns: Friend or Foe?', *Mol Neurobiol,* 54 (2017), 8071-89. DOI: 10.1007/s12035-016-0297-1.

[13] Wang X. F., Huang L. D., Yu P. P., Hu J. G., Yin L., Wang L., Xu X. M., and Lu P. H., 'Upregulation of Type I Interleukin-1 Receptor after Traumatic Spinal Cord Injury in Adult Rats', *Acta Neuropathol,* 111 (2006), 220-8. DOI: 10.1007/s00401-005-0016-x.

[14] Mukhamedshina Y. O., Akhmetzyanova E. R., Martynova E. V., Khaiboullina S. F., Galieva L. R., and Rizvanov A. A., 'Systemic and Local Cytokine Profile Following Spinal Cord Injury in Rats: A Multiplex Analysis', *Front Neurol,* 8 (2017), 581. DOI: 10.3389/fneur.2017.00581.

[15] Garlanda C., Dinarello C. A., and Mantovani A., 'The Interleukin-1 Family: Back to the Future', *Immunity,* 39 (2013), 1003-18. DOI: 10.1016/j.immuni.2013.11.010.

[16] Liu S., Xu G. Y., Johnson K. M., Echetebu C., Ye Z. S., Hulsebosch C. E.,

and McAdoo D. J., 'Regulation of Interleukin-1beta by the Interleukin-1 Receptor Antagonist in the Glutamate-Injured Spinal Cord: Endogenous Neuroprotection', *Brain Res,* 1231 (2008), 63-74. DOI: 10.1016/j.brainres.2008.07.035.

[17] Tonai T., Taketani Y., Ueda N., Nishisho T., Ohmoto Y., Sakata Y., MuraguchiM.,WadaK.,andYamamotoS., 'Possible Involvement of Interleukin-1 in Cyclooxygenase-2 Induction after Spinal Cord Injury in Rats', *J Neurochem,* 72 (1999), 302-9. DOI: 10.1046/j.1471-4159.1999.0720302.x.

[18] Sato A., Ohtaki H., Tsumuraya T., Song D., Ohara K., Asano M., Iwakura Y., Atsumi T., and Shioda S., 'Interleukin-1 Participates in the Classical and Alternative Activation of Microglia/Macrophages after Spinal Cord Injury', *J Neuroinflammation,* 9 (2012), 65. DOI: 10.1186/1742-2094-9-65.

[19] Mason J. L., Suzuki K., Chaplin D. D., and Matsushima G. K., 'Interleukin-1beta Promotes Repair of the Cns', *J Neurosci,* 21 (2001), 7046-52.

[20] Koyasu S., and Moro K., 'Type 2 Innate Immune Responses and the Natural Helper Cell', *Immunology,* 132 (2011), 475-81. DOI: 10.1111/j.1365-2567.2011.03413.x.

[21] Pomeshchik Y., Kidin I., Korhonen P., Savchenko E., JaronenM.,LehtonenS.,WojciechowskiS., Kanninen K., Koistinaho J., and Malm T., 'Interleukin-33 Treatment Reduces Secondary Injury and Improves Functional Recovery after Contusion Spinal Cord Injury', *Brain Behav Immun,* 44 (2015), 68-81. DOI: 10.1016/j.bbi.2014.08.002.

[22] Cayrol C., and Girard J. P., 'Interleukin-33 (Il-33): A Nuclear Cytokine from the Il-1 Family', *Immunol Rev,* 281 (2018), 154-68. DOI: 10.1111/imr.12619.

[23] Benczik M., and Gaffen S. L., 'The Interleukin (Il)-2 Family Cytokines: Survival and Proliferation Signaling Pathways in T Lymphocytes', *Immunol Invest,* 33 (2004), 109-42. DOI: 10.1081/imm-120030732.

[24] Nelson B. H., 'Il-2, Regulatory T Cells, and Tolerance', *J Immunol,* 172 (2004), 3983-8. DOI: 10.4049/jimmunol.172.7.3983.

[25] Malek T. R., and Castro I., 'Interleukin-2 Receptor Signaling: At the Interface between Tolerance and Immunity', *Immunity,* 33 (2010), 153-65. DOI: 10.1016/j.immuni.2010.08.004.

[26] Waters R. S., Perry J. S. A., Han S., Bielekova B., and Gedeon T., 'The Effects of Interleukin-2 on Immune Response Regulation', *Math Med Biol,* 35 (2018), 79-119. DOI: 10.1093/imammb/dqw021.

[27] Waldmann T. A., 'The Shared and Contrasting Roles of Il2 and Il15 in the Life and Death of Normal and Neoplastic Lymphocytes: Implications for Cancer Therapy', *Cancer Immunol Res,* 3 (2015), 219-27. DOI: 10.1158/2326-6066.CIR-15-0009.

[28] Gordon S., and Martinez F. O., 'Alternative Activation of Macrophages: Mechanism and Functions', *Immunity,* 32 (2010), 593-604. DOI: 10.1016/j.immuni.2010.05.007.

[29] Gadani S. P., Cronk J. C., Norris G. T., and Kipnis J., 'Il-4 in the Brain: A Cytokine to Remember', *J Immunol,* 189 (2012), 4213-9. DOI: 10.4049/jimmunol.1202246.

[30] Andrews A. L., Holloway J. W., Holgate S. T., and Davies D. E., 'Il-4 Receptor Alpha Is an Important Modulator of Il-4 and Il-13 Receptor Binding: Implications for the

Development of Therapeutic Targets', *J Immunol,* 176 (2006), 7456-61. DOI: 10.4049/jimmunol.176.12.7456.

[31] Junttila I. S., 'Tuning the Cytokine Responses: An Update on Interleukin (Il)-4 and Il-13 Receptor Complexes', *Front Immunol,* 9 (2018), 888. DOI: 10.3389/fimmu.2018.00888.

[32] Lima R., Monteiro S., Lopes J. P., Barradas P., Vasconcelos N. L., Gomes E. D., Assuncao-Silva R. C., Teixeira F. G., Morais M., Sousa N., Salgado A. J., and Silva N. A., 'Systemic Interleukin-4 Administration after Spinal Cord Injury Modulates Inflammation and Promotes Neuroprotection', *Pharmaceuticals (Basel),* 10 (2017). DOI: 10.3390/ph10040083.

[33] Francos-Quijorna I., Amo-Aparicio J., Martinez-Muriana A., and Lopez-Vales R., 'Il-4 Drives Microglia and Macrophages toward a Phenotype Conducive for Tissue Repair and Functional Recovery after Spinal Cord Injury', *Glia,* 64 (2016), 2079-92. DOI: 10.1002/glia.23041.

[34] de Vries J. E., 'The Role of Il-13 and Its Receptor in Allergy and Inflammatory Responses', *J Allergy Clin Immunol,* 102 (1998), 165-9. DOI: 10.1016/s0091-6749(98)70080-6.

[35] McCormick S. M., and Heller N. M., 'Commentary: Il-4 and Il-13 Receptors and Signaling', *Cytokine,* 75 (2015), 38-50. DOI: 10.1016/j.cyto.2015.05.023.

[36] Fichtner-Feigl Stefan, Strober Warren, Kawakami Koji, Puri Raj K., and Kitani Atsushi, 'Il-13 Signaling through the Il-13α2 Receptor Is Involved in Induction of Tgf-B1 Production and Fibrosis', *Nature Medicine,* 12 (2006), 99-106. DOI: 10.1038/nm1332.

[37] Dooley D., Lemmens E., Vangansewinkel T., Le Blon D., Hoornaert C., Ponsaerts P., and Hendrix S., 'Cell-Based Delivery of Interleukin-13 Directs Alternative Activation of Macrophages Resulting in Improved Functional Outcome after Spinal Cord Injury', *Stem Cell Reports,* 7 (2016), 1099-115. DOI: 10.1016/j.stemcr.2016.11.005.

[38] Yuan X., Wu Q., Wang P., Jing Y., Yao H., Tang Y., Han R., He W., Li Z., Zhang H., and Xiu R., 'Intraspinal Administration of Interleukin-7 Promotes Neuronal Apoptosis and Limits Functional Recovery through Jak/Stat5 Pathway Following Spinal Cord Injury', *Biochem Biophys Res Commun,* 514 (2019), 1023-29. DOI: 10.1016/j.bbrc.2019.04.159.

[39] Mazzucchelli R., and Durum S. K., 'Interleukin-7 Receptor Expression: Intelligent Design', *Nat Rev Immunol,* 7 (2007), 144-54. DOI: 10.1038/nri2023.

[40] Bao C., Wang B., Yang F., and Chen L., 'Blockade of Interleukin-7 Receptor Shapes Macrophage Alternative Activation and Promotes Functional Recovery after Spinal Cord Injury', *Neuroscience,* 371 (2018), 518-27. DOI: 10.1016/j.neuroscience.2017.10.022.

[41] Wu Z., Xue H. H., Bernard J., Zeng R., Issakov D., Bollenbacher-Reilley J., Belyakov I. M., Oh S., Berzofsky J. A., and Leonard W. J., 'The Il-15 Receptor {Alpha} Chain Cytoplasmic Domain Is Critical for Normal Il-15ralpha Function but Is Not Required for Trans-Presentation', *Blood,* 112 (2008), 4411-9. DOI: 10.1182/blood-2007-03-080697.

[42] Gomez-Nicola D., Valle-Argos B., Suardiaz M., Taylor J. S., and Nieto-Sampedro M., 'Role of Il-15 in Spinal Cord and Sciatic Nerve after Chronic Constriction Injury: Regulation of Macrophage and T-Cell Infiltration', *J Neurochem,* 107 (2008), 1741-52. DOI: 10.1111/j.1471-4159.2008.05746.x.

[43] Solaymani-Mohammadi S., Eckmann L., and Singer S. M.,

'Interleukin (Il)-21 in Inflammation and Immunity During Parasitic Diseases', *Front Cell Infect Microbiol,* 9 (2019), 401. DOI: 10.3389/fcimb.2019.00401.

[44] Fu Q., Liu Y., Liu X., Zhang Q., Chen L., Peng J., Ao J., Li Y., Wang S., Song G., Yu L., Liu J., and Zhang T., 'Engrafted Peripheral Blood-Derived Mesenchymal Stem Cells Promote Locomotive Recovery in Adult Rats after Spinal Cord Injury', *Am J Transl Res,* 9 (2017), 3950-66.

[45] Zong S., Zeng G., Fang Y., Peng J., Tao Y., Li K., and Zhao J., 'The Role of Il-17 Promotes Spinal Cord Neuroinflammation Via Activation of the Transcription Factor Stat3 after Spinal Cord Injury in the Rat', *Mediators Inflamm,* 2014 (2014), 786947. DOI: 10.1155/2014/786947.

[46] Dougan M., Dranoff G., and Dougan S. K., 'Gm-Csf, Il-3, and Il-5 Family of Cytokines: Regulators of Inflammation', *Immunity,* 50 (2019), 796-811. DOI: 10.1016/j.immuni.2019.03.022.

[47] Broughton S. E., Hercus T. R., Nero T. L., Kan W. L., Barry E. F., Dottore M., Cheung Tung Shing K. S., Morton C. J., Dhagat U., Hardy M. P., Wilson N. J., Downton M. T., Schieber C., Hughes T. P., Lopez A. F., and Parker M. W., 'A Dual Role for the N-Terminal Domain of the Il-3 Receptor in Cell Signalling', *Nat Commun,* 9 (2018), 386. DOI: 10.1038/s41467-017-02633-7.

[48] Takatsu K., 'Interleukin-5 and Il-5 Receptor in Health and Diseases', *Proc Jpn Acad Ser B Phys Biol Sci,* 87 (2011), 463-85. DOI: 10.2183/pjab.87.463.

[49] Shi L. L., Zhang N., Xie X. M., Chen Y. J., Wang R., Shen L., Zhou J. S., Hu J. G., and Lu H. Z., 'Transcriptome Profile of Rat Genes in Injured Spinal Cord at Different Stages by Rna-Sequencing', *BMC Genomics,* 18 (2017), 173. DOI: 10.1186/s12864-017-3532-x.

[50] van Leeuwen B. H., Martinson M. E., Webb G. C., and Young I. G., 'Molecular Organization of the Cytokine Gene Cluster, Involving the Human Il-3, Il-4, Il-5, and Gm-Csf Genes, on Human Chromosome 5', *Blood,* 73 (1989), 1142-8.

[51] Rose-John S., 'Interleukin-6 Family Cytokines', *Cold Spring Harb Perspect Biol,* 10 (2018). DOI: 10.1101/cshperspect.a028415.

[52] Guschin D., Rogers N., Briscoe J., Witthuhn B., Watling D., Horn F., Pellegrini S., Yasukawa K., Heinrich P., Stark G. R., and et al., 'A Major Role for the Protein Tyrosine Kinase Jak1 in the Jak/Stat Signal Transduction Pathway in Response to Interleukin-6', *EMBO J,* 14 (1995), 1421-9.

[53] Van Wagoner N. J., and Benveniste E. N., 'Interleukin-6 Expression and Regulation in Astrocytes', *J Neuroimmunol,* 100 (1999), 124-39. DOI: 10.1016/s0165-5728(99)00187-3.

[54] Pineau I., and Lacroix S., 'Proinflammatory Cytokine Synthesis in the Injured Mouse Spinal Cord: Multiphasic Expression Pattern and Identification of the Cell Types Involved', *J Comp Neurol,* 500 (2007), 267-85. DOI: 10.1002/cne.21149.

[55] Kwon B. K., Stammers A. M., Belanger L. M., Bernardo A., Chan D., Bishop C. M., Slobogean G. P., Zhang H., Umedaly H., Giffin M., Street J., Boyd M. C., Paquette S. J., Fisher C. G., and Dvorak M. F., 'Cerebrospinal Fluid Inflammatory Cytokines and Biomarkers of Injury Severity in Acute Human Spinal Cord Injury', *J Neurotrauma,* 27 (2010), 669-82. DOI: 10.1089/neu.2009.1080.

[56] Lacroix S., Chang L., Rose-John S., and Tuszynski M. H., 'Delivery of Hyper-Interleukin-6 to the Injured Spinal Cord Increases Neutrophil and Macrophage Infiltration and Inhibits Axonal Growth', *J Comp Neurol,* 454 (2002), 213-28. DOI: 10.1002/cne.10407.

[57] Nakamura M., Okada S., Toyama Y., and Okano H., 'Role of Il-6 in Spinal Cord Injury in a Mouse Model', *Clin Rev Allergy Immunol,* 28 (2005), 197-204. DOI: 10.1385/CRIAI:28:3:197.

[58] Okada S., Nakamura M., Mikami Y., Shimazaki T., Mihara M., Ohsugi Y., Iwamoto Y., Yoshizaki K., Kishimoto T., Toyama Y., and Okano H., 'Blockade of Interleukin-6 Receptor Suppresses Reactive Astrogliosis and Ameliorates Functional Recovery in Experimental Spinal Cord Injury', *J Neurosci Res,* 76 (2004), 265-76. DOI: 10.1002/jnr.20044.

[59] Guerrero A. R., Uchida K., Nakajima H., Watanabe S., Nakamura M., Johnson W. E., and Baba H., 'Blockade of Interleukin-6 Signaling Inhibits the Classic Pathway and Promotes an Alternative Pathway of Macrophage Activation after Spinal Cord Injury in Mice', *J Neuroinflammation,* 9 (2012), 40. DOI: 10.1186/1742-2094-9-40.

[60] Bonni A., Sun Y., Nadal-Vicens M., Bhatt A., Frank D. A., Rozovsky I., Stahl N., Yancopoulos G. D., and Greenberg M. E., 'Regulation of Gliogenesis in the Central Nervous System by the Jak-Stat Signaling Pathway', *Science,* 278 (1997), 477-83. DOI: 10.1126/science.278.5337.477.

[61] Klein M. A., Moller J. C., Jones L. L., Bluethmann H., Kreutzberg G. W., and Raivich G., 'Impaired Neuroglial Activation in Interleukin-6 Deficient Mice', *Glia,* 19 (1997), 227-33. DOI: 10.1002/(sici)1098-1136(199703)19:3<227::aid-glia5>3.0.co;2-w.

[62] Brunello A. G., Weissenberger J., Kappeler A., Vallan C., Peters M., Rose-John S., and Weis J., 'Astrocytic Alterations in Interleukin-6/Soluble Interleukin-6 Receptor Alpha Double-Transgenic Mice', *Am J Pathol,* 157 (2000), 1485-93. DOI: 10.1016/s0002-9440(10)64787-6.

[63] Hakkoum D., Stoppini L., and Muller D., 'Interleukin-6 Promotes Sprouting and Functional Recovery in Lesioned Organotypic Hippocampal Slice Cultures', *J Neurochem,* 100 (2007), 747-57. DOI: 10.1111/j.1471-4159.2006.04257.x.

[64] Yang G., and Tang W. Y., 'Resistance of Interleukin-6 to the Extracellular Inhibitory Environment Promotes Axonal Regeneration and Functional Recovery Following Spinal Cord Injury', *Int J Mol Med,* 39 (2017), 437-45. DOI: 10.3892/ijmm.2017.2848.

[65] Leng S. X., and Elias J. A., 'Interleukin-11 Inhibits Macrophage Interleukin-12 Production', *J Immunol,* 159 (1997), 2161-8.

[66] Trepicchio W. L., Wang L., Bozza M., and Dorner A. J., 'Il-11 Regulates Macrophage Effector Function through the Inhibition of Nuclear Factor-Kappab', *J Immunol,* 159 (1997), 5661-70.

[67] Schwertschlag U. S., Trepicchio W. L., Dykstra K. H., Keith J. C., Turner K. J., and Dorner A. J., 'Hematopoietic, Immunomodulatory and Epithelial Effects of Interleukin-11', *Leukemia,* 13 (1999), 1307-15. DOI: 10.1038/sj.leu.2401514.

[68] Zhu Z., Lee C. G., Zheng T., Chupp G., Wang J., Homer R. J., Noble P. W., Hamid Q., and Elias J. A., 'Airway Inflammation and Remodeling in Asthma. Lessons from Interleukin 11 and Interleukin 13 Transgenic Mice', *Am J Respir Crit Care Med,* 164 (2001),

S67-70. DOI: 10.1164/ajrccm.164.supplement_2.2106070.

[69] Ropeleski M. J., Tang J., Walsh-Reitz M. M., Musch M. W., and Chang E. B., 'Interleukin-11-Induced Heat Shock Protein 25 Confers Intestinal Epithelial-Specific Cytoprotection from Oxidant Stress', *Gastroenterology,* 124 (2003), 1358-68. DOI: 10.1016/s0016-5085(03)00282-8.

[70] Gurfein B. T., Zhang Y., Lopez C. B., Argaw A. T., Zameer A., Moran T. M., and John G. R., 'Il-11 Regulates Autoimmune Demyelination', *J Immunol,* 183 (2009), 4229-40. DOI: 10.4049/jimmunol.0900622.

[71] Cho N., Nguyen D. H., Satkunendrarajah K., Branch D. R., and Fehlings M. G., 'Evaluating the Role of Il-11, a Novel Cytokine in the Il-6 Family, in a Mouse Model of Spinal Cord Injury', *J Neuroinflammation,* 9 (2012), 134. DOI: 10.1186/1742-2094-9-134.

[72] Maheshwari A., Janssens K., Bogie J., Van Den Haute C., Struys T., Lambrichts I., Baekelandt V., Stinissen P., Hendriks J. J., Slaets H., and Hellings N., 'Local Overexpression of Interleukin-11 in the Central Nervous System Limits Demyelination and Enhances Remyelination', *Mediators Inflamm,* 2013 (2013), 685317. DOI: 10.1155/2013/685317.

[73] Russo R. C., Garcia C. C., Teixeira M. M., and Amaral F. A., 'The Cxcl8/Il-8 Chemokine Family and Its Receptors in Inflammatory Diseases', *Expert Rev Clin Immunol,* 10 (2014), 593-619. DOI: 10.1586/1744666X.2014.894886.

[74] McTigue D. M., Tani M., Krivacic K., Chernosky A., Kelner G. S., Maciejewski D., Maki R., Ransohoff R. M., and Stokes B. T., 'SelectiveChemokineMrnaAccumulation in the Rat Spinal Cord after Contusion Injury', *J Neurosci Res,* 53 (1998), 368-76. DOI: 10.1002/(SICI)1097-4547(19980801)53:3<368::AID-JNR11>3.0.CO;2-1.

[75] Stammers A. T., Liu J., and Kwon B. K., 'Expression of Inflammatory Cytokines Following Acute Spinal Cord Injury in a Rodent Model', *J Neurosci Res,* 90 (2012), 782-90. DOI: 10.1002/jnr.22820.

[76] Taylor A. R., Welsh C. J., Young C., Spoor E., Kerwin S. C., Griffin J. F., Levine G. J., Cohen N. D., and Levine J. M., 'Cerebrospinal Fluid Inflammatory Cytokines and Chemokines in Naturally Occurring Canine Spinal Cord Injury', *J Neurotrauma,* 31 (2014), 1561-9. DOI: 10.1089/neu.2014.3405.

[77] Kwon B. K., Streijger F., Fallah N., Noonan V. K., Belanger L. M., Ritchie L., Paquette S. J., Ailon T., Boyd M. C., Street J., Fisher C. G., and Dvorak M. F., 'Cerebrospinal Fluid Biomarkers to Stratify Injury Severity and Predict Outcome in Human Traumatic Spinal Cord Injury', *J Neurotrauma,* 34 (2017), 567-80. DOI: 10.1089/neu.2016.4435.

[78] Mueller C. A., Schluesener H. J., Conrad S., Pietsch T., and Schwab J. M., 'Spinal Cord Injury-Induced Expression of the Immune-Regulatory Chemokine Interleukin-16 Caused by Activated Microglia/Macrophages and Cd8+ Cells', *J Neurosurg Spine,* 4 (2006), 233-40. DOI: 10.3171/spi.2006.4.3.233.

[79] Hermann E., Darcissac E., Idziorek T., Capron A., and Bahr G. M., 'Recombinant Interleukin-16 Selectively Modulates Surface Receptor Expression and Cytokine Release in Macrophages and Dendritic Cells', *Immunology,* 97 (1999), 241-8. DOI: 10.1046/j.1365-2567.1999.00786.x.

[80] Jana M., and Pahan K., 'Il-12 P40 Homodimer, but Not Il-12 P70, Induces the Expression of Il-16 in Microglia

and Macrophages', *Mol Immunol,* 46 (2009), 773-83. DOI: 10.1016/j.molimm.2008.10.033.

[81] Fickenscher H., Hor S., Kupers H., Knappe A., Wittmann S., and Sticht H., 'The Interleukin-10 Family of Cytokines', *Trends Immunol,* 23 (2002), 89-96. DOI: 10.1016/s1471-4906(01)02149-4.

[82] Couper K. N., Blount D. G., and Riley E. M., 'Il-10: The Master Regulator of Immunity to Infection', *J Immunol,* 180 (2008), 5771-7. DOI: 10.4049/jimmunol.180.9.5771.

[83] Shouval D. S., Ouahed J., Biswas A., Goettel J. A., Horwitz B. H., Klein C., Muise A. M., and Snapper S. B., 'Interleukin 10 Receptor Signaling: Master Regulator of Intestinal Mucosal Homeostasis in Mice and Humans', *Adv Immunol,* 122 (2014), 177-210. DOI: 10.1016/B978-0-12-800267-4.00005-5.

[84] Hellenbrand D. J., Reichl K. A., Travis B. J., Filipp M. E., Khalil A. S., Pulito D. J., Gavigan A. V., Maginot E. R., Arnold M. T., Adler A. G., Murphy W. L., and Hanna A. S., 'Sustained Interleukin-10 Delivery Reduces Inflammation and Improves Motor Function after Spinal Cord Injury', *J Neuroinflammation,* 16 (2019), 93. DOI: 10.1186/s12974-019-1479-3.

[85] Commins S., Steinke J. W., and Borish L., 'The Extended Il-10 Superfamily: Il-10, Il-19, Il-20, Il-22, Il-24, Il-26, Il-28, and Il-29', *J Allergy Clin Immunol,* 121 (2008), 1108-11. DOI: 10.1016/j.jaci.2008.02.026.

[86] Horiuchi H., Parajuli B., Wang Y., Azuma Y. T., Mizuno T., Takeuchi H., and Suzumura A., 'Interleukin-19 Acts as a Negative Autocrine Regulator of Activated Microglia', *PLoS One,* 10 (2015), e0118640. DOI: 10.1371/journal.pone.0118640.

[87] Guo J., Wang H., Li L., Yuan Y., Shi X., and Hou S., 'Treatment with Il-19 Improves Locomotor Functional Recovery after Contusion Trauma to the Spinal Cord', *Br J Pharmacol,* 175 (2018), 2611-21. DOI: 10.1111/bph.14193.

[88] Lee J. S., Hsu Y. H., Chiu Y. S., Jou I. M., and Chang M. S., 'Anti-Il-20 Antibody Improved Motor Function and Reduced Glial Scar Formation after Traumatic Spinal Cord Injury in Rats', *J Neuroinflammation,* 17 (2020), 156. DOI: 10.1186/s12974-020-01814-4.

[89] Sun L., He C., Nair L., Yeung J., and Egwuagu C. E., 'Interleukin 12 (Il-12) Family Cytokines: Role in Immune Pathogenesis and Treatment of Cns Autoimmune Disease', *Cytokine,* 75 (2015), 249-55. DOI: 10.1016/j.cyto.2015.01.030.

[90] Liu J., Cao S., Kim S., Chung E. Y., Homma Y., Guan X., Jimenez V., and Ma X., 'Interleukin-12: An Update on Its Immunological Activities, Signaling and Regulation of Gene Expression', *Curr Immunol Rev,* 1 (2005), 119-37. DOI: 10.2174/1573395054065115.

[91] Hamza T., Barnett J. B., and Li B., 'Interleukin 12 a Key Immunoregulatory Cytokine in Infection Applications', *Int J Mol Sci,* 11 (2010), 789-806. DOI: 10.3390/ijms11030789.

[92] Yaguchi M., Ohta S., Toyama Y., Kawakami Y., and Toda M., 'Functional Recovery after Spinal Cord Injury in Mice through Activation of Microglia and Dendritic Cells after Il-12 Administration', *J Neurosci Res,* 86 (2008), 1972-80. DOI: 10.1002/jnr.21658.

[93] Hymowitz S. G., Filvaroff E. H., Yin J. P., Lee J., Cai L., Risser P., Maruoka M., Mao W., Foster J., Kelley R. F., Pan G., Gurney A. L., de Vos A. M., and Starovasnik M. A., 'Il-17s Adopt a Cystine Knot Fold: Structure and Activity of a Novel Cytokine, Il-17f, and Implications for Receptor Binding', *EMBO J,* 20 (2001), 5332-41. DOI: 10.1093/emboj/20.19.5332.

[94] Pappu R., Ramirez-Carrozzi V., and Sambandam A., 'The Interleukin-17 Cytokine Family: Critical Players in Host Defence and Inflammatory Diseases', *Immunology,* 134 (2011), 8-16. DOI: 10.1111/j.1365-2567.2011.03465.x.

[95] Onishi R. M., and Gaffen S. L., 'Interleukin-17 and Its Target Genes: Mechanisms of Interleukin-17 Function in Disease', *Immunology,* 129 (2010), 311-21. DOI: 10.1111/j.1365-2567.2009.03240.x.

[96] Hill F., Kim C. F., Gorrie C. A., and Moalem-Taylor G., 'Interleukin-17 Deficiency Improves Locomotor Recovery and Tissue Sparing after Spinal Cord Contusion Injury in Mice', *Neurosci Lett,* 487 (2011), 363-7. DOI: 10.1016/j.neulet.2010.10.057.

[97] Dooley D., Lemmens E., Ponsaerts P., and Hendrix S., 'Interleukin-25 Is Detrimental for Recovery after Spinal Cord Injury in Mice', *J Neuroinflammation,* 13 (2016), 101. DOI: 10.1186/s12974-016-0566-y.